CHROMATOPIA
AN ILLUSTRATED
HISTORY OF COLOUR

DAVID COLES

WITH PHOTOGRAPHY BY ADRIAN LANDER

FOR LOUISE

INTRODUCTION
vii

A LIFE LIVED IN COLOUR
ix

GLOSSARY
xiv

THE BASICS OF COLOUR
1
BLUE
PURPLE
RED
ORANGE
YELLOW
GREEN

MANUFACTURING PIGMENTS
177
LEAD WHITE
CARMINE LAKE
ULTRAMARINE
MADDER LAKE

ARTISTS' COLOUR
187

ENDNOTES
222

ACKNOWLEDGEMENTS
223

I

THE FIRST COLOURS
14
OCHRES
CHALK WHITE
LAMP BLACK
BONE WHITE
BONE BLACK

II

COLOUR IN THE TIME OF THE ANCIENTS
26
EGYPTIAN BLUE
ORPIMENT
REALGAR
WOAD

VI

DYES, LAKES + PINKES
92
ARZICA
BRAZILWOOD
LOGWOOD
STIL DE GRAIN
MADDER LAKE
COCHINEAL

THE FARMERS OF COCHINEAL
106

VII

MYSTERIOUS COLOURS
110
INDIAN YELLOW
GAMBOGE
MUMMY BROWN

III

COLOUR + THE CLASSICAL WORLD

36

LEAD WHITE
TYRIAN PURPLE
INDIGO
MALACHITE
AZURITE
RED LEAD
VERDIGRIS
CHRYSOCOLLA

IV

MEDIEVAL COLOURS

54

LAC
VINE BLACK
KERMES
DRAGON'S BLOOD
LAPIS LAZULI
PEACH BLACK
LEAD TIN YELLOW
VERMILION
SMALT
SAFFRON
BLUE VERDITER
GRAPHITE
NAPLES YELLOW

V

WRITING INKS

82

GALL INK
BISTRE
SEPIA
WALNUT

VIII

THE EXPLOSION OF COLOURS

118

PRUSSIAN BLUE
LEAD CHROMATES
EMERALD GREEN
COBALT
POTTER'S PINK
ULTRAMARINE
CADMIUM
CERULEAN BLUE
MANGANESE

IX

A BRAVE NEW WORLD OF COLOUR

138

MARS COLOURS
ZINC WHITE
TITANIUM WHITE

SYNTHETIC CHEMISTRY OF THE MODERN WORLD

146

X

THE SCIENCE OF MODERN COLOUR

152

FLUORESCENCE
PHOTOLUMINESCENCE
YInMn BLUE
VANTABLACK

MAKING COLOUR VISCERAL: HOW PAINT IS MADE

162

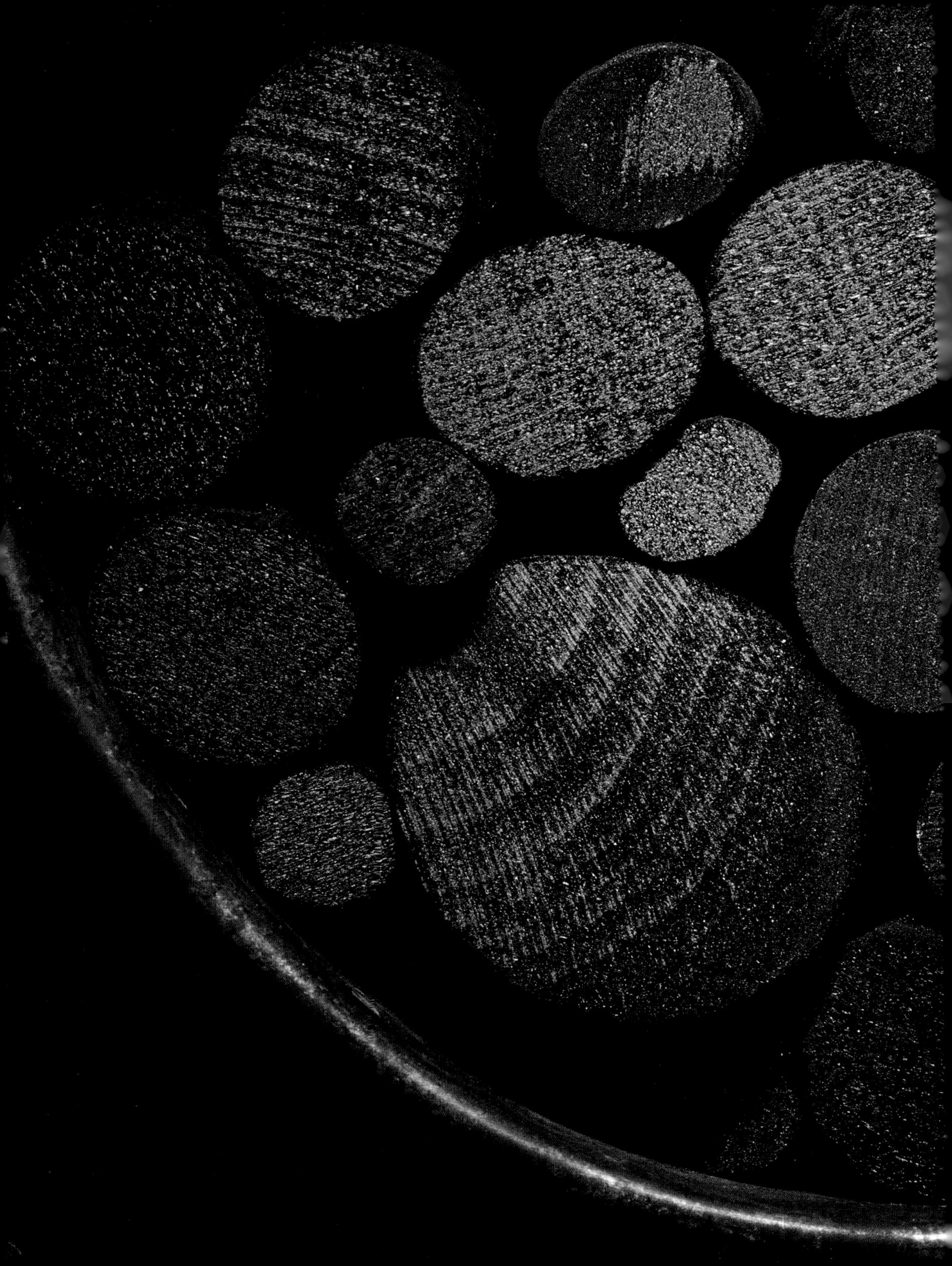

INTRODUCTION

Since the beginnings of human existence, colour has played an integral role in the way we describe the world around us. From the first pigments drawn out of the earth to the earliest civilisations creating vibrant pigments from base materials when no such colour was naturally available, colour has always played an important part in recording the history of mankind. Our ingenuity in fashioning colour out of dirt led to the magnificence of medieval manuscripts, the art of the Renaissance, and the expressions of 20th-century Modernism. Modern science has brought us colours of unimaginable vibrancy and pigments that deny their third-dimension or glow in complete darkness.

Spanning thousands of years, the history of pigments and their incredible journeys across the globe are revealed within these pages. From colours dug from the ground beneath us to others so expensive only kings and popes could afford them, they have all brought beauty to the world.

From the ancient world to the present day, tales of dragons and beetles, alchemy and poisons, slaves and pirates are brought to life in this book on the creation of colour.

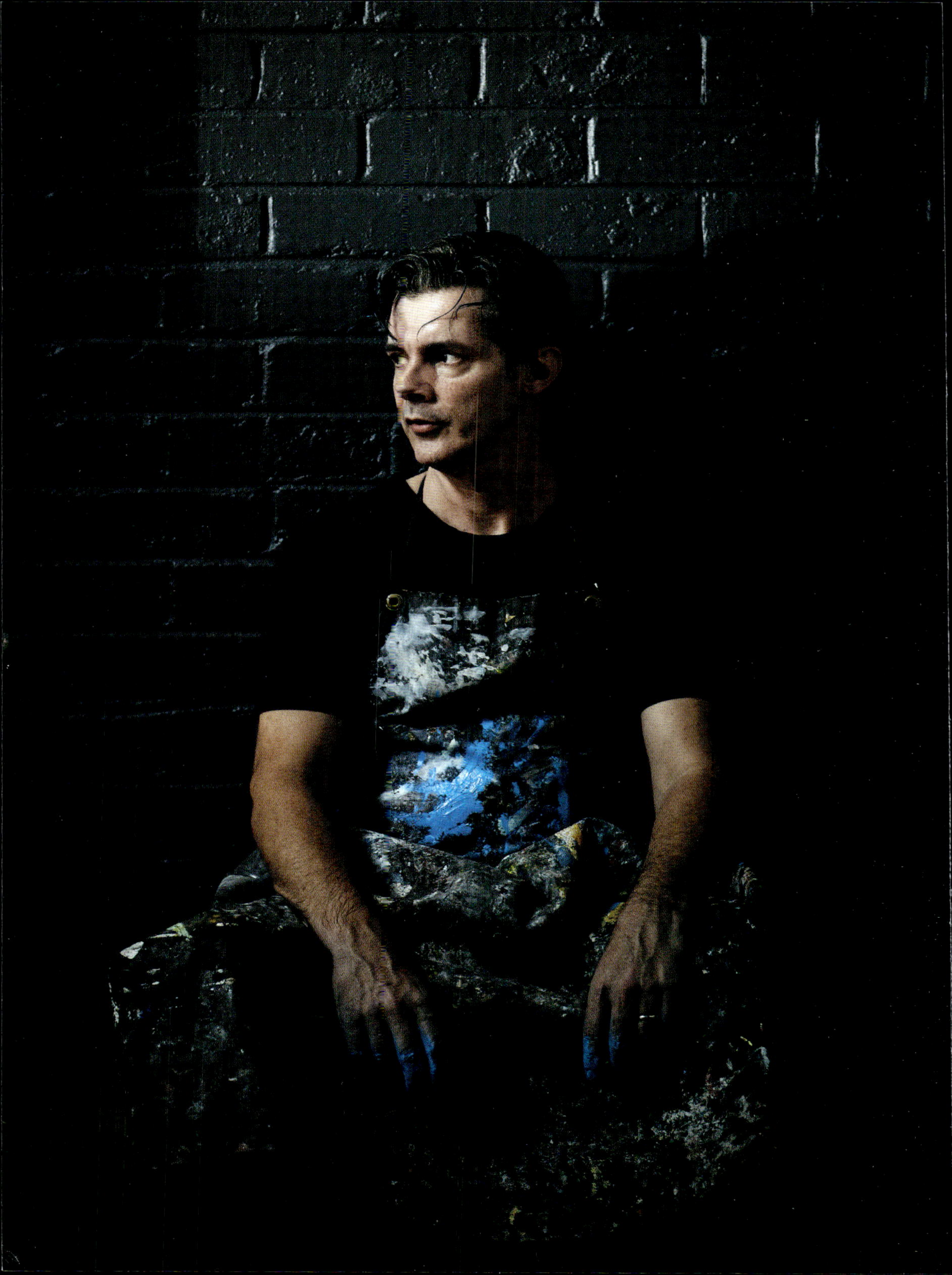

A LIFE LIVED IN COLOUR

I have spent my whole life within the world of colour. Looking back, it could be mistaken for destiny but how I got to this place in time is a tale of serendipity, family legacy, chance meetings, wrong turns and a tenacious pursuit of the alchemical transformation of dirt into colour. In the process, I founded a paint-making company that has become one of the world's most respected makers of artists' oil paints.

My father was an advertising illustrator; the house was always full of paints, paper, markers and pencils. During school holidays, my brother and I would take the train with him up to his London studio, part of a ramshackle early-Victorian five-storey building split up into a labyrinth of studios for artists, writers, calligraphers and agents. That building is no longer standing; it is one of many that were pulled down during the redevelopment of Covent Garden. During one of these visits, a most significant event occurred. I was taken to Cornelissen & Son, the famous art materials store, then still in Great Queen Street where it had stood since 1863. Walking in, I was overwhelmed by the smell of oil-primed canvas that had accreted to the interior of the store over decades, and the unctuous aroma of painting oils and balsams. Arrayed in a giant cabinet of solid black timber were giant glass jars of pure pigments, natural resins, gums and waxes. Each jar had a label, describing its contents – *gamboge*, *Japan wax*, *sandarac*, *terre-verte*, *copaiba balsam*, *gum elemi* – tempting me into a world of stories of their exotic origins.

Back in my country town, I began working in the family art store. I had already started to paint and the knowledge I learnt in the shop began to inform my art practice. Around this time I was given a set of pigments, and those tiny jars – sparkling, like powdered jewels – set me on my path to an obsession with colour. With support from my parents I applied myself to becoming an artist, and at eighteen started my studies in painting at Bristol Art College under a traditional system of studying composition and colour.

An integral part of our training was the preparation and manufacture of the artist's materials: canvas, paint, mediums and varnishes. This was the early 1980s, painting was back in favour, and students around me were rediscovering the paints and materials abandoned by previous generations. Some experimented with the ancient technique of encaustic, painting with pigmented molten wax. Others tried their hand at distemper, hand-making their paint from warm solutions of rabbit-skin glue, chalk and pigment. All this experimentation, and the encouragement of our lecturers, left us with the realisation that we were part of a 'golden chain' of history that could be traced back all the way through the centuries to the Renaissance and beyond.

After college I moved to London and ended up getting a job, over a decade after my first visit, at Cornelissen's. Here I truly began my apprenticeship in the history of colour. As one of the most venerable art materials stores in the world, it had incredibly high standards for knowledge of the materials we sold and how they should be mastered. Working with like-minded staff heightened my immense curiosity about the materials of artists and traditional craftsmen. Many of the specialist tradespeople who frequented the store used techniques going back centuries. Through them, I entered a world of craft lore, where it was understood that the physical could hold metaphysical and magical qualities.

After work, evenings at home were spent making concoctions from 18th-century recipes – cooking, dissolving, filtrating and synthesising raw materials and pigments to hand-make paints, mediums and grounds. It was eternally fascinating. That was the point when I realised my future was to work with colour, from its raw ingredients through to beautifully crafted paint.

My discovery of Australia was accidental. In 1989 I decided to take a six-month holiday, travelling through India and trekking in the recently opened Annapurna mountain ranges in Nepal. On returning to my base at Kathmandu, I was handed an opportunity to fly on to Thailand which I grabbed. By then, the money was truly running out and I had to make a decision: continue or return home. Realising I wanted to keep exploring, I applied for a working holiday visa at the Australian embassy, cancelled my return airline tickets to the UK, took local buses all the way to Singapore and bought a one-way ticket to Darwin. I ended up in a different world of vast skies, extreme heat and friendly locals.

In 1992 I emigrated to Australia, determined to build a business on my knowledge and passion for colour. Founded with the ridiculously small capital of $2000, Langridge Artist Colours initially supplied the materials I knew best – artist-quality pigments and my own range of oil mediums – to artists and friends within the Melbourne art world.

For the first twelve years, the original factory was in Langridge Street (from which the company derives its name), in Collingwood. At this time I didn't have enough money to buy the equipment to manufacture oil paints, so every bit of profit was set aside to go towards the purchase of a paint-making mill.

In 1999 I inherited a small double-roll mill, the most basic of paint-making machines. Four years later, having saved enough, I was able to afford a small triple-roll mill and begin paint-making in earnest. There are, unsurprisingly, no vocational courses on artist paint-making and I was too young and ambitious to want an apprenticeship with one of the handful of paint-makers in the world. Instead, I threw myself into what ended up being years of trial and error and self-education – thousands of hours hunched over a triple-roll mill attempting to make an oil paint worthy of my ambition. Our dioxazine violet – a modern, crazily intense deep cool purple – took ten laborious hours to mill. The end result, however, was paint so full of pigment that the purple looked almost black as it was squeezed from the tube.

Things became a lot easier for me when I was lucky enough to purchase, at a ridiculously knocked-down price, a large, fifty-year-old Lehmann-made mill. With granite rollers, it was built for the processing of chocolate. Luckily for us, it can also make the best oil paint in the world.

In 2004 Langridge moved to a larger factory in Yarraville in Melbourne's inner west where I continued to refine the oil paint formulations. The following year, we finally released our oil paint to the world.

In recent years, we have created our own unique colours at Langridge. The first of these, Brilliant Blue, attempts to replicate the light-filled blue of the Australian sky, which is vastly different to that of Europe or America. The searing sunlight here excites all colours it touches, creating a high chromatic vibration that is immediate and modern. Alongside modern single-pigment colours I have expanded our range to include Video Green, Brilliant Magenta and Neon Orange, to replicate the modern world built under the intense sunlight of Australia.

After almost forty years working with colour I'm still amazed at its power to excite emotions. Whether it is the discovery of a pigment's history (some are as old as the ages), or the electrifying sensation of seeing a new modern pigment for the first time, this excitement has never left me.

Now, on the occasion when I am still able to get into the studio, I spend many hours simply luxuriating in mixing two or more colours together to see the result. Generally this is for no particular purpose other than to experience the visual alchemy of individual colours interacting to create the most unexpected results.

Ultimately, paint is just a tool that helps the artist create a work of art. But I will admit that my heart leaps a little when I see a painting and recognise one of my paints. There can be nothing more rewarding than that.

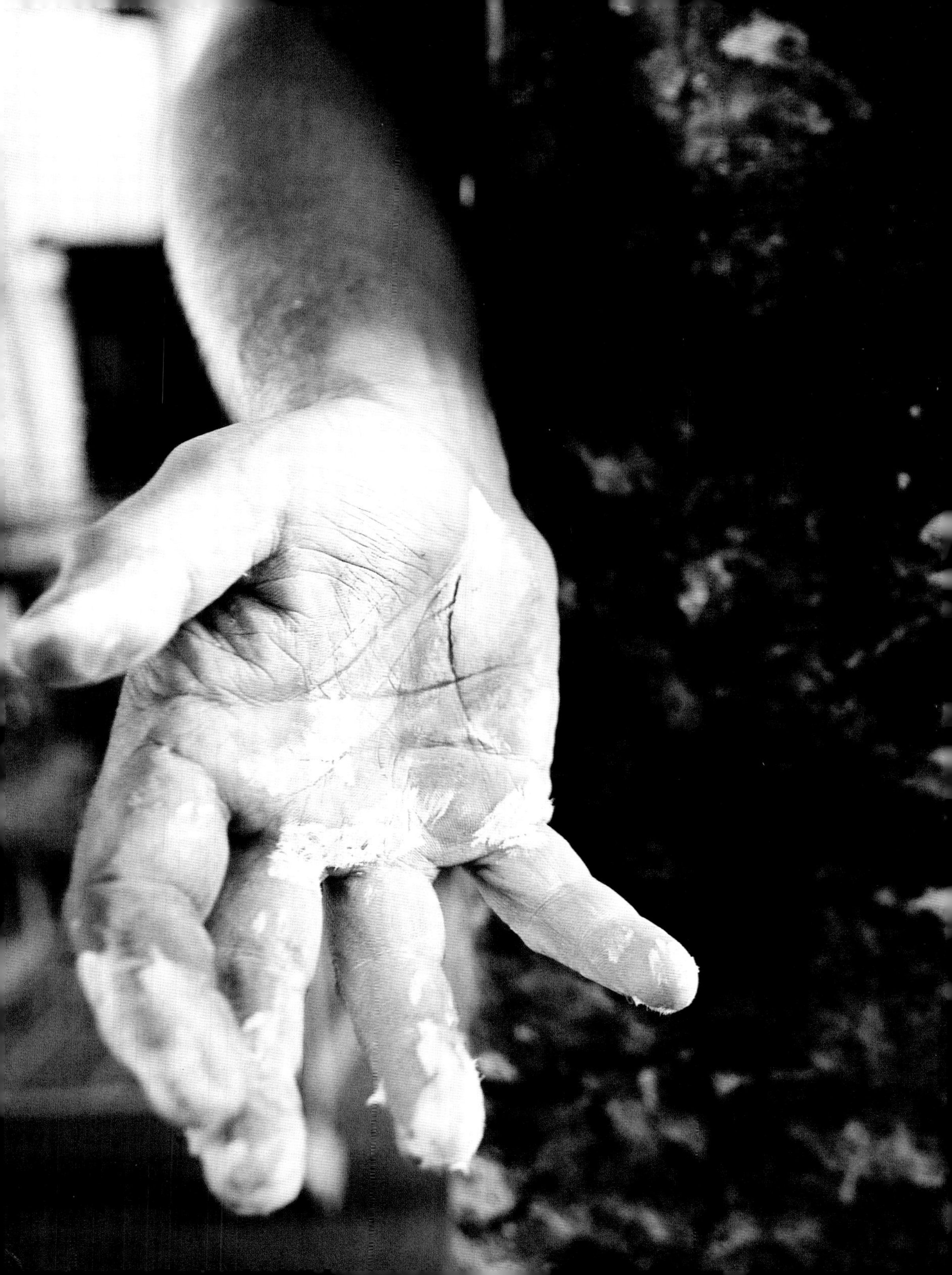

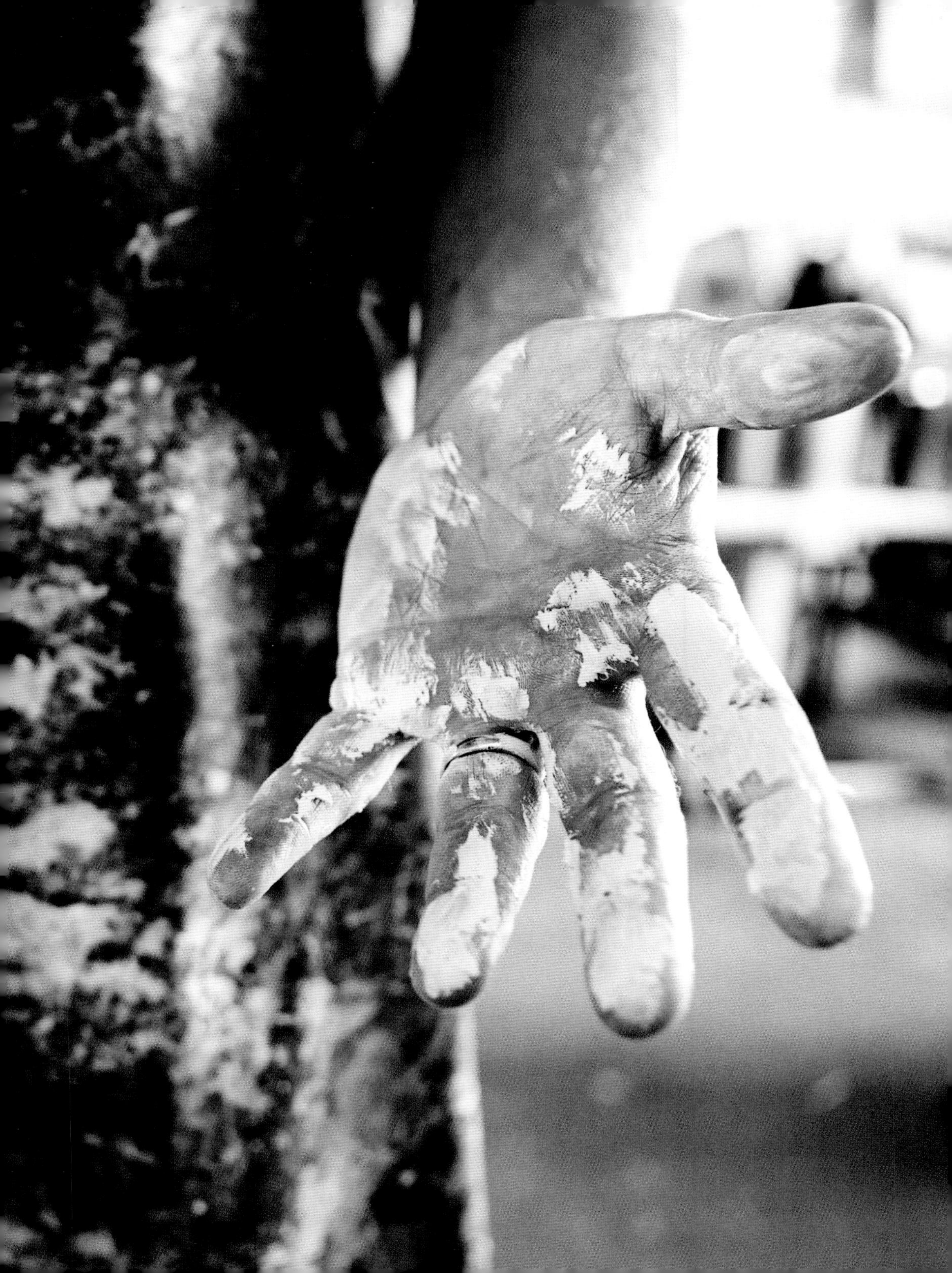

GLOSSARY

To help you understand some of the specific names and historical definitions, here are explanations for the more frequently used terms in this book.

ADDITIVE COLOUR
Colour principle where colours are divisions of wavelengths of white light. Adding the primary colours together results in unified white light.

ALKALI
An ionic salt with a pH greater than 7.0. Concentrated solutions are caustic, causing chemical burns.

ALUM
Also known as aluminium sulphate, this salt is added to alkali solutions to create aluminium hydroxide. It is used to prepare lake pigments by converting dyes into insoluble pigments.

ANCIENT WORLD
The period of history from the first civilisations of Mesopotamia, Egypt and India to the start of the classical world of Greece.

ARTIFICIAL
Manufactured, rather than occurring naturally.

BINDER
A substance that holds pigment particles together to make paint. The individual characteristics of the binding agent influence paint qualities such as drying time, sheen and colour saturation.

BODY COLOUR
The opaque colour of an unadulterated or undiluted pigment or paint.

CHROMA
The intensity or saturation of a colour.

CLASSICAL WORLD
The period of Greek civilisation from the early 8th century BCE through to the end of the Western Roman Empire in the late 5th century CE.

DISTEMPER
Paint made from a warm solution of pigment bound with animal glue.

DYE
A colourant that is fully soluble, allowing it to stain other materials.

EGG TEMPERA
A paint made of pigment bound with egg yolk.

FUGITIVE COLOUR
A pigment or dye that, when exposed to light, moisture, or adverse temperature, loses its colour or changes in appearance.

GESSO
A mixture of chalk and animal glue made from the hides or bones of rabbits.

GLAIR
A clear binder made from beaten egg whites, used for making manuscript inks and paints.

GLAZE
A coloured or transparent paint coating that allows colours beneath it to be seen.

GUM ARABIC
The water-soluble sap of the acacia tree, used as a binder for watercolours and gouache.

HIDING POWER
A measure of the opacity of a paint or pigment and its ability to obscure the surface beneath it.

HUE
The quality that distinguishes one colour from another, such as a red from a yellow, green, blue or purple.

INK
A highly fluid coloured solution for writing or drawing, made from a dye or pigment and usually applied with a pen or brush.

INORGANIC PIGMENT
A pigment that has a mineral origin, such as metal-based pigments made from iron, cadmium, cobalt or lead.

LAKE PIGMENT
A pigment created by chemically reacting a dye with aluminium hydroxide.

LIGHTFAST COLOUR
The property of pigment to resist fading or discolouration from exposure to sunlight.

LUMINESCENCE
The process by which a substance absorbs energy from a source that does not involve heat, such as light or a biochemical reaction, and then emits that energy as visible light.

NATRON
Naturally occurring sodium carbonate, found in salt lake beds. Historically important as an alkali used to manufacture soap, glass and pigments.

ORGANIC PIGMENT
An insoluble pigment that comes from a vegetable or animal source, or from carbon-based compounds derived from chemically refined petroleum oil.

PHOTOLUMINESCENCE
The phenomenon by which a material absorbs energy from photons of ultraviolet light and then emits it as a visible light, often for long periods of time after the energy source is no longer present.

PIGMENT
An insoluble, individual particle of colour.

PINKE
The historical term for 'lake pigment'. This term does not describe the colour.

POTASH
Also known as potassium carbonate, this alkali is used to manufacture lake pigments. Refers to the historical production method of soaking ashes in a pot of water.

PRIMER
A prepared coating that is applied to a surface to prepare it for painting.

SILVERPOINT
A drawing technique using a fine rod of silver applied to a slightly abrasive surface.

SUBTRACTIVE COLOUR
The term used for the colour principle based on physical colour (i.e. pigments), whereby the three primary colours reduce in chroma and value when mixed together.

SYNTHETIC
The creation of a new material by the reaction of two or more elements.

TANNIN
Gallic acid derived from vegetal material. This creates a black-brown dye.

THE CRAFTSMAN'S HANDBOOK
Written by the artist Cennino Cennini in the 15th century. A complete manual of the materials and techniques of medieval artistic painting.

VALUE
The relative lightness or darkness of a colour.

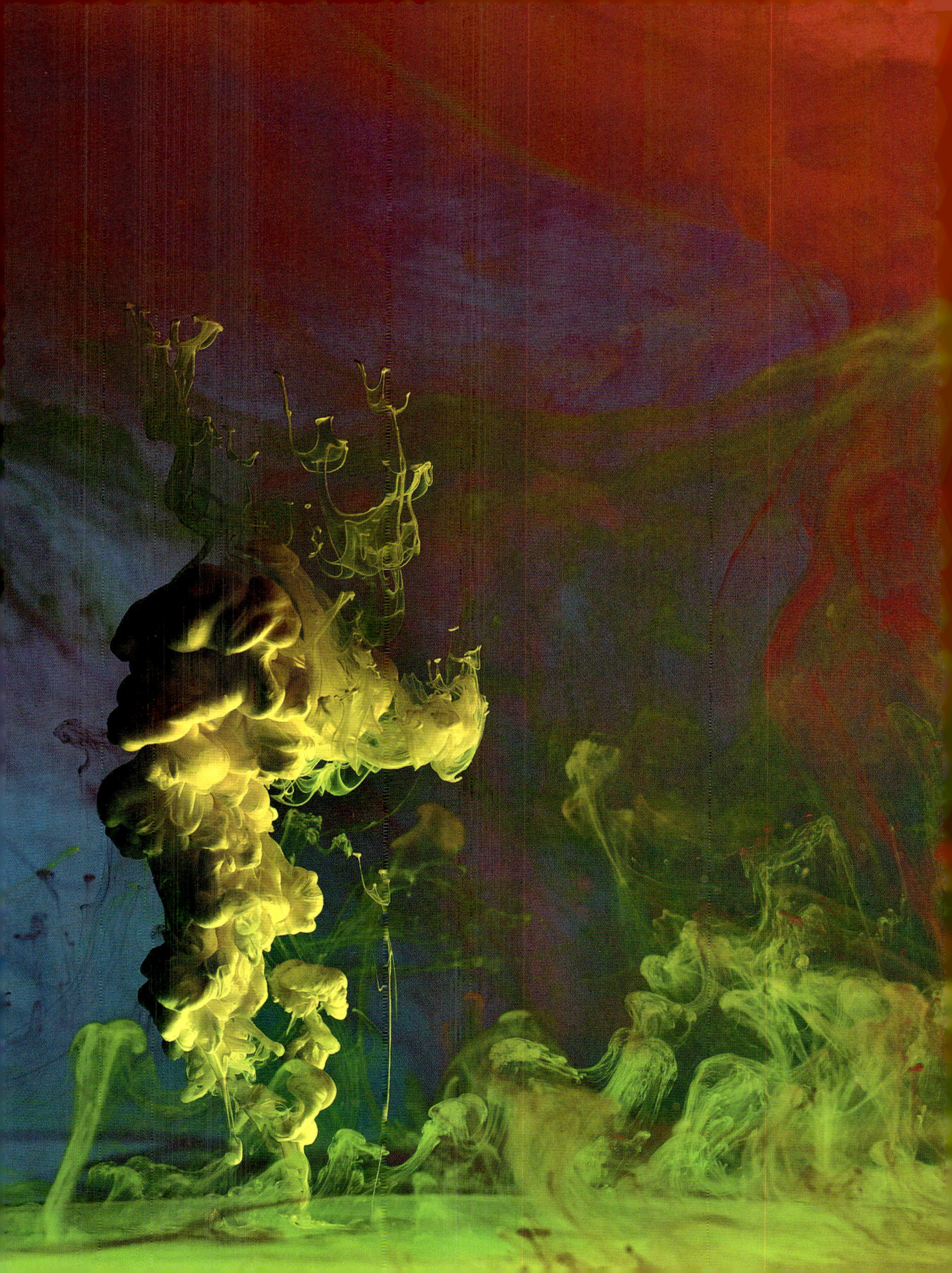

THE BASICS OF COLOUR

To understand how to navigate colour it is important to know the rules of physical colour, commonly called the subtractive colour model. In this model, there are three primary colours: red, blue and yellow. From these, virtually all other colours can be mixed. By adding two primaries together we end up with the secondary colours; red and blue make purple, blue and yellow make green, yellow and red create orange. The addition of more colours creates tertiary colours, but every time more colours are added, the purity of colour drops until eventually we end up with browns and greys.

Most of the pigments of history were chromatically weak and artists who wanted to keep bright colours in their paintings were loath to mix them. The history of pigments is full of technological advances, each age creating brighter and purer colours that artists have hungrily adopted.

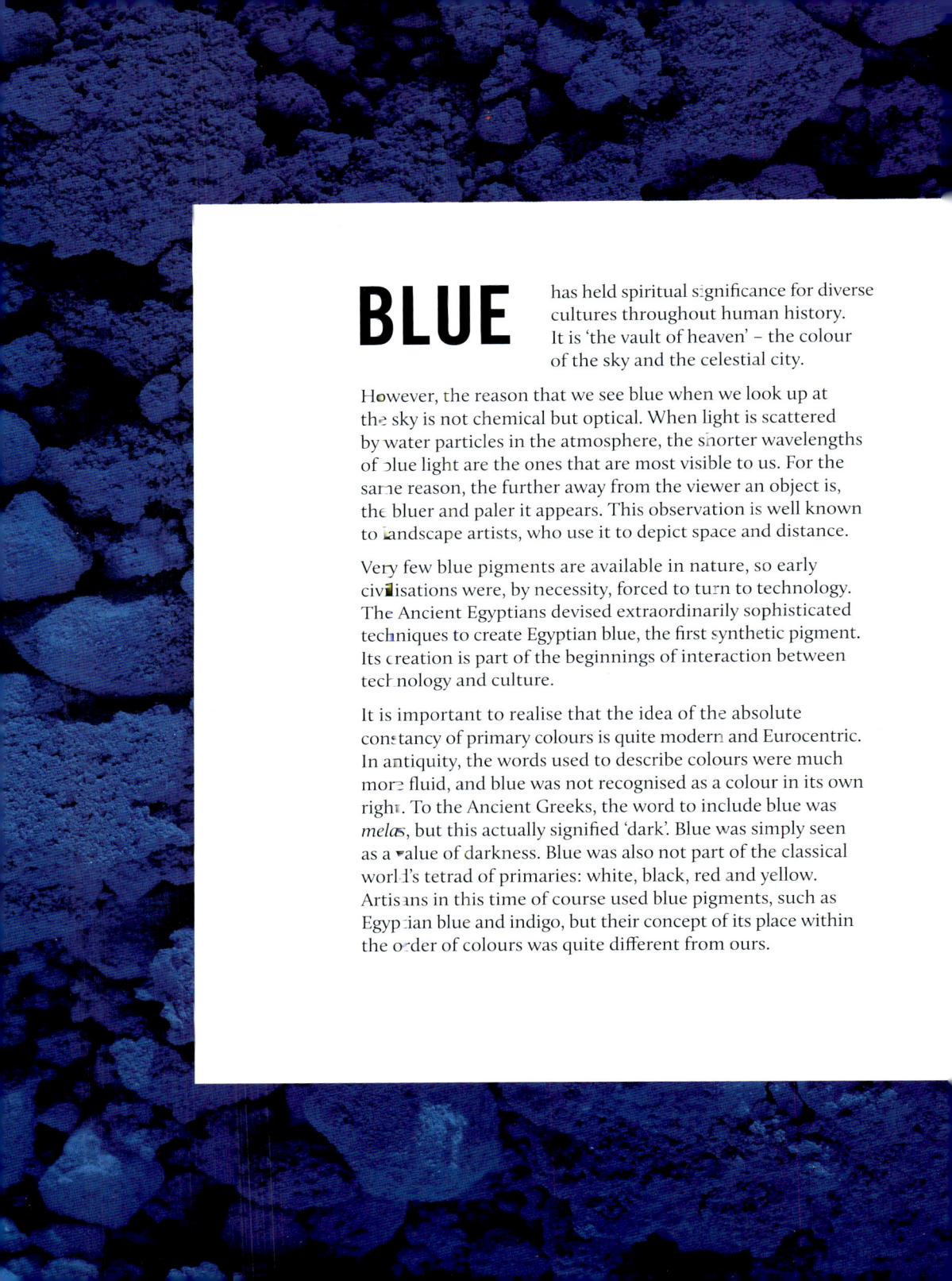

BLUE

has held spiritual significance for diverse cultures throughout human history. It is 'the vault of heaven' – the colour of the sky and the celestial city.

However, the reason that we see blue when we look up at the sky is not chemical but optical. When light is scattered by water particles in the atmosphere, the shorter wavelengths of blue light are the ones that are most visible to us. For the same reason, the further away from the viewer an object is, the bluer and paler it appears. This observation is well known to landscape artists, who use it to depict space and distance.

Very few blue pigments are available in nature, so early civilisations were, by necessity, forced to turn to technology. The Ancient Egyptians devised extraordinarily sophisticated techniques to create Egyptian blue, the first synthetic pigment. Its creation is part of the beginnings of interaction between technology and culture.

It is important to realise that the idea of the absolute constancy of primary colours is quite modern and Eurocentric. In antiquity, the words used to describe colours were much more fluid, and blue was not recognised as a colour in its own right. To the Ancient Greeks, the word to include blue was *melas*, but this actually signified 'dark'. Blue was simply seen as a value of darkness. Blue was also not part of the classical world's tetrad of primaries: white, black, red and yellow. Artisans in this time of course used blue pigments, such as Egyptian blue and indigo, but their concept of its place within the order of colours was quite different from ours.

Used sparingly throughout the Middle Ages, due to the relative rarity or dullness of most available pigments, blue finally moved to centre stage when ultramarine arrived in Europe in the 13th century. Derived from the semi-precious stone lapis lazuli, ultramarine cost more than gold itself. Its use was a display of wealth, prestige and devotion. In paintings of the Virgin Mary, her blue mantle emphasises her position as the queen of heaven, an intermediary between God and humanity.

The tonal values of many pigments, especially the blues, change when they are bound in different mediums. Ultramarine is a brilliant luminous blue in egg tempera but it becomes dark in oil. When linseed oil was introduced as a binder for paints in the early 15th century, artists were forced to add white to restore its brilliance. This reversed a long-held prejudice against mixing pigments dating back to the ancient world.

The scarcity of blue pigments ceased dramatically at the beginning of the 18th century with the introduction of Prussian blue, followed quickly by cobalt blue, cerulean blue and synthetic ultramarine. These technical advances in pigment manufacture gave blue a central role in the modern palette. Its apotheosis came in the late 1950s, when Yves Klein created monochromatic paintings using his signature colour, International Klein Blue. His works bring the philosophic idea of blue as 'atmosphere' full circle from the beginnings of civilisation to the present day.

PURPLE

has always been associated with royalty. Of all the primary and secondary colours, it has the most enduring singular meaning. Its luxurious richness flows strongly through its history.

The term 'purple', from the Latin *purpura*, is derived from the Ancient Greek *porphyra*. This was the name for the Tyrian purple dye that was manufactured in classical antiquity. Extracted from a Mediterranean sea snail, it was the most precious of ancient dyestuffs.

The colour purple is often described by its relationship to its adjacent primaries: red and blue. The Chinese saying, 'It is so red, it is purple', means that a dark, intense red will shift to purple (red purple brings luck and fame). In Afghanistan, the purest grade of lapis lazuli is called *surpar* (red feather), which describes a blue that verges on the violet at the very heart of a fire. Throughout the ancient and medieval world, the term *purpura* could mean a shade of dark red, crimson or purple, and had strong associations with blood. The Roman author Pliny remarked, 'The Tyrian colour is most appreciated when it is the colour of clotted blood, dark by reflected and brilliant by transmitted light'.

Purple is the most extravagant of colours. It is traditionally associated with royalty and nobility because Tyrian purple was only affordable for the elite. During the Roman Empire, 'The Purple' meant the Roman emperor, a term derived from the colour of his robe. Byzantine empresses gave birth in the Purple Chamber and *porphyrogenitus* (born to the purple) described an emperor who gained the throne by inheritance and not force.

The Mayans, Aztecs and Incas had their own mollusc-based purple dye, which was also associated with religious ceremonies and royalty. In Japan, purple signifies position and wealth and to the Chinese it was the colour of the aristocracy. Within the Christian church, it is the colour of repentance, contrition and suffering, symbolising the purple robe that Jesus was made to wear before the crucifixion.

Like blue, purple reflects many of our notions of the atmosphere, light and space. In the 19th century, the Impressionists replaced blacks and browns with purples to depict shadows. Their love for the newly invented cobalt and manganese violets led to accusations from critics that they suffered from 'violettomania'.

For most of human civilisation, the rarity of pure purple gave rise to its symbolic nature, but in 1856 Henry Perkin accidentally invented the very first synthetic dye, which he initially called Tyrian purple (it was later changed to 'mauveine'). It was an astute marketing ploy to associate a cheaply produced, artificial tincture with the most expensive dye in all of history.

RED is not easy to categorise. It has carried many different meanings across cultures and periods in history. From the bright scarlet of vermilion to the cool crimson of carmine lake, its history is as varied and bold as the colour itself.

As red is the colour of blood, it has historically been associated with sacrifice, violence and courage. In Roman mythology, it was the colour of Mars, the god of war. Roman soldiers wore red tunics, their generals wore scarlet cloaks, and they celebrated victories by painting their bodies entirely in red. Earlier, Egyptians coloured themselves with red ochre during celebrations, as did many ancient cultures, including the Indigenous peoples of Australia.

But red can also signify charity, fertility and love. Brides in Ancient Rome wore a red shawl, Chinese brides wear red wedding gowns and, in Greece and other Balkan nations, brides still wear red veils. In many Asian countries, red is the colour of good fortune and happiness.

To the Ancient Greeks, red was an intermediate colour that sat halfway between light and dark. Unlike us, they decoded colour using tonal values. Many civilisations' understanding of an individual colour's position was driven by what it is was made of, where it came from, how it behaved as a pigment, and how it was used in their society. This idea continued well into the 18th century and explains much of the confusion and contradiction about red throughout history. In the classical world, red and green were 'joined' (due to having equal darkness). In the Middle Ages, 'scarlet' meant a quality fabric and in the 17th century 'pink' was the term used for lake pigments of many different colours. Only later did the words 'scarlet' and 'pink' become attached to specific red hues.

The red colour of the earth comes from naturally occurring oxides of iron, and ochre was humanity's first red. Depending on its source, red ochre could be vivid or dull. To obtain a primary red, we turned to technology. Red lead, invented by the Ancient Greeks, is one of the earliest manufactured pigments and it was the dominant red in European painting until the introduction of vermilion in the 8th century.

Vermilion is a product of alchemy. The raw materials – sulphur and mercury – were considered the parents of all metals and were essential to the transmutation of base materials to gold. Vermilion is so bright that it completely transformed the artists' palette. All other colours had to compete for attention, which spurred the creation of stronger blues and yellows. The sumptuous reds of red lead and vermilion, however, were opaque and tended towards orange.

The use of insects, such as kermes, lac and cochineal, and vegetal sources of madder roots led to lake pigments that offered artists transparent crimsons that transformed in oil paint to glazes of jewel-like intensity.

Red was adopted by the Catholic church as a symbol for the blood of Christ and the Christian martyrs. Monarchs then appropriated it as a sign of authority, and it was used by the rich mercantile classes to indicate status and, in the 20th century, by the socialist movement to signify revolution. This boldest of primary colours has always been contrary: anger and love, warmth and cold, poverty and wealth.

ORANGE

wasn't always called orange. In fact, the word we now use for the colour comes from the fruit that was introduced to Europe from Asia by Portuguese merchants in the late 15th century. Before that, the colour was referred to either as a hue of red or yellow. In Old English, it was *geoluhread*, which means 'yellow-red'. The modern term, derived from the Sanskrit *narcnga*, wasn't used to describe the colour until the middle of the 16th century.

In painting and traditional colour theory, orange is a secondary colour created by mixing red and yellow together. Throughout history, although the colour could be created from these two primaries, there have also been pigments that we would now call orange. The first orange pigments were natural ochres that sat between yellow and red ochres. Like most earth pigments, they are relatively dull as they contain many impurities that lower their chromatic intensity.

Very early on in human civilisation, however, brighter hues became available. Realgar, an arsenic-based mineral that yields a rich orange pigment, was used throughout antiquity. Its relative scarcity led the Ancient Romans to develop one of the first synthetic pigments – red lead. Although the name suggests a primary colour, it was available as an orange-red, and the Egyptians considered it to be a shade of yellow.

Across many cultures, the perception of orange as a colour in its own right was allied to spiritual transmutation. In Hinduism, orange represents fire and is a metaphor for the inner transformation that is experienced by swamis donning orange robes. Orange also signified change for the Confucians of Ancient China. In both China and India, the colour took its name not from the fruit but from saffron. In Buddhism, orange (or, more precisely, saffron) is the colour of illumination, the highest state of perfection. To many in the western world, orange is a mixture of the energy associated with red and the happiness connected to yellow. It is related to warmth, creativity, change and activity.

In the 19th century, artists took advantage of new, more chromatically true hues to raise the brightness of their palette. In 1872, Claude Monet painted *Impression, Sunrise*, which featured a tiny but vivid chrome orange sun against a misty riverscape of blues. The placement of orange against its complementary colour of blue creates a searing visual effect.

The later introduction of the more chemically stable cadmium orange further entrenched orange within the modern palette and the arrival of the 20th century heralded the organic synthetic pigments that are as luminous and saturated as the flaming sun.

YELLOW

was used by ancient cultures to replicate and harness the divine power of the sun. The sun is one of the oldest and most central symbols to humanity, worshipped in many cultures around the world. Objects and paintings that were coloured with yellow dyes, pigments or the most precious yellow of all, gold, were not just symbols of devotion. They could also intervene on behalf of the believer in everyday life.

To the Ancient Egyptians, yellow (*khenet*) represented perfection. Gold (*newb*) represented the flesh of the gods and was used to decorate the sarcophagi of the pharaohs. To the Greeks, yellow was one of the four primary colours and, in their mythology, the sun-god Helios wore a yellow robe while riding his golden chariot across the heavens. To the early Christians, yellow was associated with the Pope and the golden keys to the kingdom of heaven. In Byzantine churches, gold-covered walls reflected the sun and bathed the visitor in the radiant yellow light of the divine.

Yet yellow also has many conflicting associations. It is the colour of sunshine, optimism, creativity and enlightenment, but it also represents cowardice, duplicity and avarice. In classical Greece, prostitutes wore saffron-dyed clothes, and in Rome they dyed their hair yellow. In the later Christian church, yellow became associated with Judas Iscariot and denoted heretics. In the Middle Ages, non-Christians were marked with the colour yellow. In 16th-century Spain, those accused of heresy were compelled to come before the Inquisition dressed in a yellow cape of treason. In the

20th century, yellow was revived as a symbol of exclusion by the Nazis, who forced Jewish people to wear the yellow Star of David on their clothing.

Yellow ochre is one of the first pigments used by humans. The ancient world sourced the deadly yellow mineral orpiment and refined gold from the earth, while the Romans ingeniously manufactured lead-based yellows. In the Middle Ages, Europeans used plants for their yellow juices, employed alchemy to produce lead tin yellow and imported exotic Indian yellow and gamboge. Modern chemistry led to an explosion of colour through the introduction of chrome yellow, cadmium yellow, cobalt yellow and the synthetic azo dyes and pigments.

Curiously, there are no dark yellows. When black is added to blue or red, they become darker version of themselves, but when yellow is mixed with black it shifts to green.

Just as sunlight illuminates the gloom, yellow is the most luminous colour in the spectrum. Because of this luminosity, yellow radiates from its surroundings towards the human eye and most immediately captures our attention.

GREEN

is a secondary colour in the subtractive colour model of painting. It is made from a mixture of yellow and blue. However, in the additive colour model of light, it is one of the primary colours, sitting proudly alongside red and blue to create all other colours.

Nature is green. The Ancient Greek word for green is *chlorós*, from which we derive the term 'chlorophyll' for the pigment in plants that allows them to absorb energy from light through photosynthesis. The colour represents growth and fertility, and the word is used to describe things that are young and fresh, like newly cut timber or apprentice workers. The Latin word for green is *viridis*. Poetically the Romans used the term *viridus*, 'having green old age', to describe the age between forty-five and sixty.

To the Ancient Egyptians, green was the colour of fresh growth, new life and resurrection. It also represented healing and protective powers, and powdered green gemstones like emerald were used as eye ointments. This medical belief continued until the Middle Ages. Green is the sacred colour in Islam. It is the colour of the banner of Mohammed, and the paradise that awaits.

Most of human history has been lived within a verdant pastoral landscape, but there were very few natural green pigments to describe it. The first were green earths, but the deposits were not widespread and its use was limited. The greens derived from plants are not lightfast so artists often resorted to mixing blues and yellows to make the colours of foliage and grasses.

For centuries, the only true green pigments came from copper-bearing minerals such as malachite and chrysocolla. Green and blue are both formed from these ores and the distinction between them has often been blurred. Thus, green and blue have always danced a graceful and lyrical duet. Verdigris, the manufactured blue-green rust of copper, allowed greater access to the colour but its tendency to corrupt and turn dark meant that the colour was as short-lived as the green leaves of spring.

Like all colours, green has many contradictory meanings. In Buddhism, verdant green means life, but pale green depicts the kingdom of death. In the western world, it is both a symbol of youth and growth as well as the livid colour of disease, sickness and death. Green can also be toxic. In the late 18th century, new copper-based pigments were manufactured using arsenic and their brilliant colours poisoned thousands of innocents.

It was not until the 19th century that artists finally had access to readily available safe, stable and potent greens in the form of viridian and cobalt green. The 20th century saw the introduction of phthalo green. Its gem-like intensity restored the purity of green's association with life and vitality.

I.

THE
FIRST
COLO

URS

THE FIRST COLOURS

OCHRES

OCHRES WERE THE VERY FIRST
PIGMENTS USED BY HUMANS.

The oldest human artworks still in existence are vivid depictions of animals, humans and spirits that were created using ochres. There is evidence of their use as far back as 250,000 years ago. Ancient ochre artworks are found all over the world, from the earliest cultures of India and Australia to the famous cave paintings of Lascaux in France.

Naturally occurring iron-containing ochres of the earth provide a wide range of yellow, red and brown colours. The natural mineral could be collected or dug-up and then simply ground against a harder rock and water added to make fluid. Later civilisations refined this process to include washing the ochre of impurities, drying and then grinding to a fine powder.

Yellow ochres are an impure form of iron oxide called limonite. They can also be roasted to produce other hues by placing on a fire or in an oven. A moderate heat turns the yellow to orange; stronger heat makes the colour turn red. These roasted red ochres are often called 'burnt' (for example, burnt sienna). Naturally occurring red ochres are richer in anhydrous iron oxide called haematite. They also vary widely in shade, hue and transparency.

There are many earth pigments whose specific colour comes from natural mineral admixtures. The pigments known as 'umber' contain iron plus manganese oxide that lends a greenish hue. Iron-oxide-free earths are not strictly ochres, but it is important to include them here as their use alongside the true ochres is significant throughout history; white earths from pipe-clay, black earths of manganese and the light green pigment *terre-verte* (green earth) from mineral celadonite.

THE FIRST COLOURS

CHALK WHITE

EARLY HUMANS ADDED CHALK
TO THE OCHRE PALETTE.

Whether used alone or added to ochres to brighten them, the use of chalk expanded the range of colours. Readily available and easy to grind into a powder, it had no competition until the invention of the lead whites by the Ancient Greeks.

Chalk is a soft, white mineral called calcite or calcium carbonate. It is formed from the fossilised remains of microscopic phytoplankton algae. Chalk deposits are often very thick and extensive. The most famous of these hail from the soaring white cliffs of the southern English coastline.

Chalk in its many guises – English whiting, Bologna chalk and Champagne chalk (the different names reflect the place of origin) – has also been used since the early Renaissance to make a traditional gesso ground. Gesso is a bright white plaster surface that was applied to complex multiple timber panels for the construction of church altarpieces. Gesso unified the object and gave a very smooth priming coat to which gold leaf or egg tempera paint could be applied. With the rise of oil painting in the 15th century, artists moved from solid timber as a paint surface to flexible woven linen canvasses. Inflexible gesso was dropped in favour of supple oil primers coloured with lead white, as chalk turns almost transparent when added to linseed oil. The use of chalk as a white pigment in painting also declined for the same reason.

Since then, chalk has been used mostly as an extender for other pigments and as the bulk content of soft pastels. Because it can be manufactured cheaply, most chalk sold today is not natural, but made by a process of chemical precipitation to regulate its performance for industrial applications.

THE FIRST COLOURS

LAMP BLACK

LAMP BLACK HAS BEEN AROUND
SINCE PREHISTORIC TIMES.

This lightfast, permanent, opaque blue-black pigment was used by the Ancient Egyptians more than 4000 years ago for painting tombs and murals. They preferred its fineness and deep black colour to the grey-black of charcoal.

As you'd expect from its name, lamp black is made by collecting soot from beeswax candles or lamps that burn tallow, resin or oil. The basic process is simple: carbon is deposited on a cold surface that is suspended over an oily flame.

The manufacturing methods have been refined over thousands of years and many raw materials have been experimented with. The legendary Ancient Greek artist Apelles is credited with inventing a lamp black called *atramentum elephantinum* by collecting the soot of burnt ivory tusks. The Romans made lamp black by burning the dregs of wine – apparently the best wines produced a blue-black colour that was equal to indigo.

Throughout history, lamp black has been employed to make writing and drawing inks. It has an extremely fine grain and does not need further grinding. At the same time as the first Egyptian pharaohs, the Chinese combined lamp black and animal glue to create Chinese ink. Later, the raw materials were traded from India and the English renamed it Indian ink.

In recent years, the production of true lamp black has been generally replaced by the burning of acetylene gas rather than oils to create a purer carbon black.

THE FIRST COLOURS

BONE WHITE

THE CREATION OF BONE WHITE
IS AS LITERAL AS IT SOUNDS.

Bone white was made by burning bones in open fires until all the organic material burned away and the bone was turned to ash. There is evidence of bone white's use since Neolithic times, making it one of the very first pigment colours created by transformation.

Although almost any kind of bone will work, historical recipes were very particular about recommending the use of certain bones. Cennino Cennini's 15th-century manual, *The Craftsman's Handbook*, instructs:

Take bone from the second joints and wings of fowls, or of a capon, and the older they are the better. Just as you find them under the dining-table, put them in the fire; and when you see that they have turned whiter than ashes, draw them out, and grind them well in the porphyry.[1]

The antlers of the male European red deer, known as a hart, were used to make a bone white called hartshorn white. Unlike horns, antlers are made of bone. Deer shed them every winter, so they were collected and prepared like other bones.

Bone white is mainly composed of a mixture of calcium phosphate and calcium carbonate. The ground pigment is a gritty whitish-grey powder. Due to its slightly abrasive quality, bone white was often used in the preparation of paper for silverpoint drawings. Mixed with warm liquid rabbit-skin glue, it was painted on as a priming coat. A paste of water and bone white could also be used as a polish for metal sculptures and silverware.

THE FIRST COLOURS

BONE BLACK

BURNING BONES TO CREATE
PIGMENTS IS AN ANCIENT PRACTICE.

Like bone white, bone black is made by putting fragments of animal bones into a crucible and surrounding them with blazing coals. However, to prevent the bone turning to ash, the vessel is covered to stop air from getting in. Exposure to intense heat in the absence of oxygen turns the bones into carbonised char. After the char has cooled, it is pulverised in a mortar and pestle.

Found in prehistoric, Egyptian, Greek and Roman art, bone black was used throughout the Middle Ages and Renaissance. The 16th-century Dutch painters, in particular, made vivid use of its dense physical nature to depict garments.

Medieval artisans refined the process by washing the pigment with water, filtering it and grinding it even finer on a stone slab. This increased the intensity of the colour. Depending on the quality of preparation and the type of bone used, a variety of black hues can be created, all the way from a blue-black through to brown.

Up until the Middle Ages, bone black was not used as widely as charcoal black, probably because of the hardness of the raw material and the difficulty of grinding it down to a fine powder.

Bone black is still in use today, but often sold under the name ivory black. Genuine ivory black was made by charring waste ivory pieces and the supply was always limited because of the scarcity of ivory. However, with the outlawing of the ivory trade, the pigment now sold under this name is actually a bone black with added carbon content to increase its darkness.

II. COLOUR IN THE TIME OF THE ANCIEN[T]

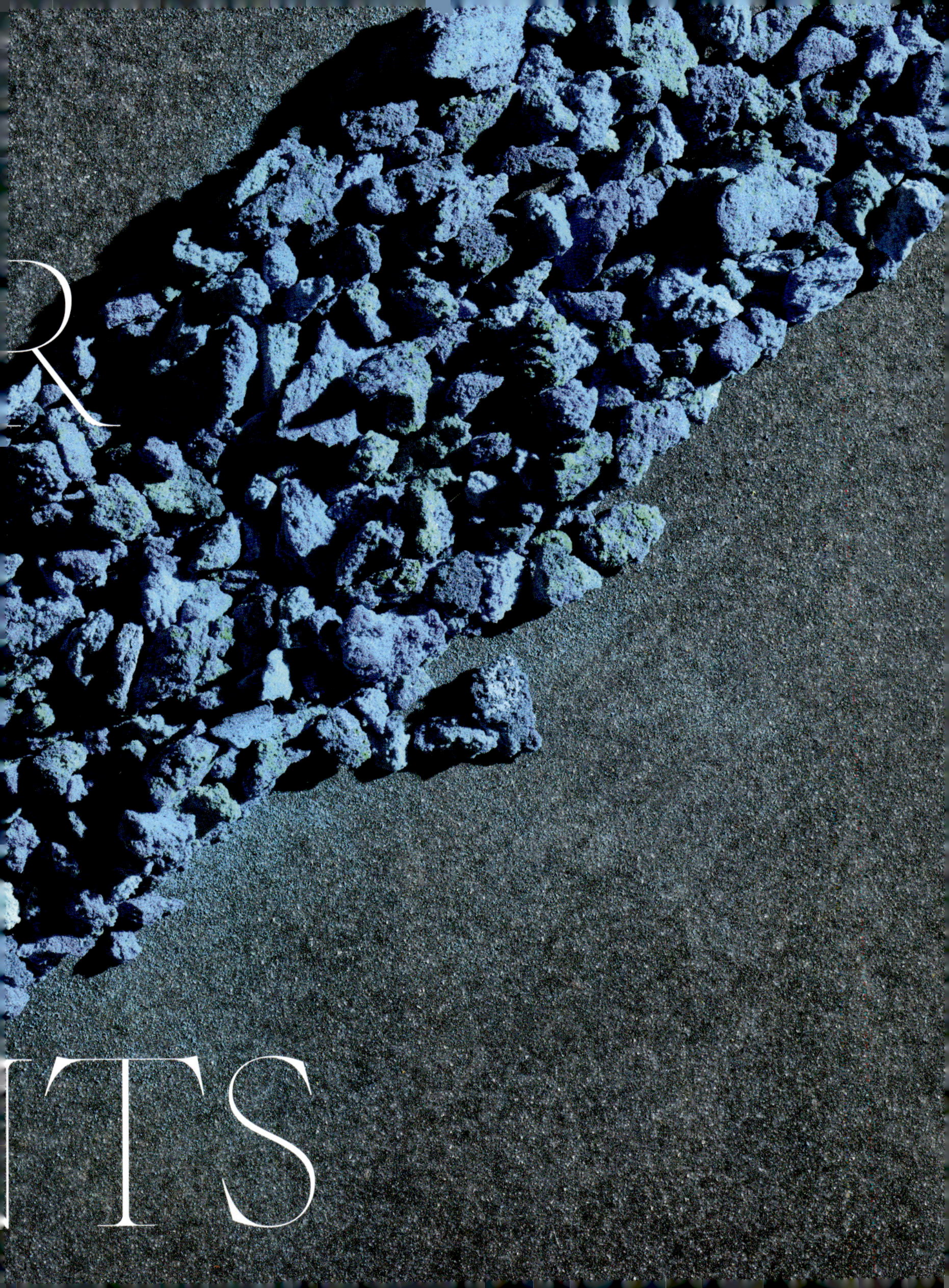
R
ITS

COLOUR IN THE TIME OF THE ANCIENTS

EGYPTIAN BLUE

THIS WAS THE FIRST SYNTHETICALLY PRODUCED COLOUR.

Invented at around the same time as the Great Pyramids were being built, Egyptian blue's creation dates back about 5000 years. The Ancient Egyptians believed blue was the colour of the heavens and because of the rarity of naturally occurring blue minerals like azurite and lapis lazuli, they devised a way to manufacture the colour themselves.

Egyptian blue was not produced by blind chance: it was created with precision. Made by heating lime, copper, silica and natron, the pigment's invention was a development of the ceramic glaze processes. The Egyptians controlled the firing of the raw materials with amazing accuracy, holding their kilns at a crucial temperature close to 830°C.

The famous crown of Queen Nefertiti owes its colour to Egyptian blue and the pigment was used extensively for painting murals, sculptures and sarcophagi. It spread from Egypt to Mesopotamia, Greece and the outer reaches of the Roman Empire and was used at the palace at Knossos, in Pompeii and on Roman wall paintings. Known to the Romans as *caeruleum* (from which the colour cerulean derives its name), it was widely used throughout the Classical Age, but the knowledge of how to make it was lost with the fall of the Roman Empire.

Discoveries made by Napoleon's 1798 Egyptian expedition led to further investigation of Egyptian blue; and eventually, in the 1880s, the chemical composition of the pigment was identified and the manufacturing process was recreated.

COLOUR IN THE TIME OF THE ANCIENTS

ORPIMENT

ORPIMENT WAS THE CLOSEST
IMITATION TO GOLD.

Its Latin name is *auripigmentum* (gold paint) and in the classical world, it was believed that this resemblance had deeper alchemical roots. It was even said that the Roman emperor Caligula could extract gold from the mineral.

In fact, orpiment carries a much more dangerous substance. It is a highly toxic sulphide of arsenic. The Persian word *zarnikh* (gold-coloured) became *arsenikon* in Greek and then *arrhenicum* in Latin, from which the English word 'arsenic' is derived. The Romans were well aware of orpiment's poisonous nature and used slave labour to mine it. For the unlucky slaves this was, in essence, a death sentence.

Orpiment was used in Ancient Egypt as a cosmetic, taking its place in history alongside other deadly pigments used in makeup. It was used in painting for centuries throughout Persia and Asia, but in Europe, because of the dominance of lead-based yellows, it was most often employed in manuscripts.

A manufactured version, known as king's yellow, was available from the 17th century. The name is believed to come from Arabic alchemy, which described orpiment and realgar as the 'two kings'.

Both the naturally occurring and synthetic versions of orpiment were incompatible with other commonly used pigments, particularly lead-based pigments like flake white, and copper-based pigments like verdigris and malachite. It was infamous for turning them black. With the introduction in the 19th century of the more chemically inert and less toxic cadmium yellow, orpiment fell out of usage.

COLOUR IN THE TIME OF THE ANCIENTS

REALGAR

THIS PIGMENT IS AS DEADLY
AS IT IS BEAUTIFUL.

Known as the 'ruby of arsenic', realgar is extremely toxic. The red crystals of the mineral yield a rich orange pigment, but it is made of arsenic disulphide. Realgar is found in the same deposits as the yellow, arsenic-containing mineral orpiment. It mostly occurs in geothermal fissures near hot springs, and the word 'realgar' comes from *rahj-al-ghar* (Arabic for 'powder of the mine').

In limited use in Ancient Egypt, artists from Mesopotamia through to India and the Far East favoured realgar. Alongside orpiment, it was a significant item of the pigment trade in Ancient Rome. The Romans called it *sandarach* and used it as an orange-red pigment for painting. In China it was called 'masculine yellow', as opposed to the 'feminine yellow' of orpiment.

Although realgar's orange hue was beguiling, it had more sinister uses. Throughout the Middle Ages it was employed as a rat poison and in China it was sprinkled around houses to repel snakes and insects.

It was used minimally in European painting, probably because another orange pigment, red lead, was easier to work with. Additionally, realgar is also not particularly stable. When it is exposed to strong light, it turns into the yellow mineral pararealgar. Like orpiment, realgar reacts with lead-based and copper-based pigments, and it can also deteriorate badly in oil paint films, resulting in ruptures and cracking.

The 16th-century Italian painter, Titian, was one of the few exponents of realgar. Its use declined in the 18th century as less toxic and more stable pigments became available.

COLOUR IN THE TIME OF THE ANCIENTS

WOAD

WOAD WAS WIDELY USED AS A DYE IN
EUROPE AS EARLY AS THE STONE AGE.

Ancient Britons covered their bodies with woad to face the Roman legions and it is said that they struck fear into Julius Caesar himself.

The first part of the woad-making process involved taking fresh leaves of the woad plant, *Isatis tinctoria*, grinding them to a pulp, rolling them into balls the size of large apples and leaving them to dry in the sun. They could then be stored and used at a later date. Like indigo, the dye is extracted by fermentation. Traditional recipes specify that the plant be soaked in urine under the heat of the sun and trampled for three days. After that, the remaining liquid is a yellowish colour.

The indigo molecule is the blue colourant in woad. The magical quality of indigo is that the distinctive blue colour only develops after the textiles are removed from the dye bath and exposed to air. During the dyeing process, a scum called florey, known as the flower of woad, also develops on the surface. This was skimmed off and dried so it could be used separately as a paint colour.

The fermentation process releases large quantities of ammonia. Far worse, however, is that the plant depletes the soil that it grows in, leaving an infertile wasteland in its wake. Laws were passed in medieval Europe to curb this devastation.

Although indigo was known since Imperial Rome, the more colour-intense Indian indigo was not readily available in the west until commercial quantities were imported at the beginning of the 17th century. It supplanted woad, and production rapidly declined as a result.

III.
COLOR IN THE GLASS WORLD

UR +

ICAL
D

COLOUR + THE CLASSICAL WORLD

LEAD WHITE

THIS IS THE GREATEST – AND THE
CRUELLEST – OF THE WHITES.

Lead white has been in continuous production for at least 2000 years. It is basic lead carbonate, formed by the reaction of lead with vapours of vinegar (acetic acid) and carbon dioxide. The manufacturing process in the 19th century was relatively unchanged from that used in the classical world.

The introduction of the 'stack' method in the 16th century refined the procedure to improve the pigment's quality. Specially constructed clay pots were divided into two separate chambers, one for the lead and the other for vinegar. Dozens of pots were lined up, covered then surrounded with large quantities of warm manure that produced carbon dioxide as well as the heat needed to accelerate the reaction. Multiple layers of pots were stacked and covered with more manure. The room was sealed and left closed for up to 90 days. This alchemical magic is the result of the corrosion of the metal, which transforms grey lead into the purest white. The process turns the solid metal into flakes, which is why lead white is also known as flake white.

For hundreds of years, this dense silver-white has been the most important pigment for artists. It is hard to imagine the history of art without it. However, it has one major defect: lead is so poisonous that prolonged exposure will kill you. This is not such an issue for artists, whose contact with it in paint is limited, but for workers in lead white factories the symptoms of poisoning included headaches, memory loss, abdominal pain and eventually death. In the late 19th century, safer synthetic whites superseded lead white. Zinc white was its first competitor and then, in the 20th century, the introduction of titanium white almost completely replaced lead white's commercial use.

COLOUR + THE CLASSICAL WORLD

TYRIAN PURPLE

THIS PRESTIGIOUS PIGMENT COMES
FROM A PREDATORY SEA SNAIL.

Tyrian purple is extracted from *Bolinus brandaris*, a mollusc native to the ancient Phoenician city of Tyre (Phoenicia means 'land of purple') in what is now Lebanon. The production of Tyrian purple goes back at least 3500 years and Greek legends tell us it was discovered by Hercules, who, on seeing his dog's purple-stained mouth, realised the colour came from the snail the dog had just chewed.

Each snail yielded just one drop of dye. A single ounce required the sacrifice of around 250,000 snails. During the height of its production in the Roman Empire, the putrid stench of millions of decomposing snails meant its manufacture was banished to the edge of town. The resulting piles of shells still litter the eastern shore of the Mediterranean.

Tyrian purple was strictly reserved for those of high rank. In Imperial Rome, prohibitive rules became so severe that only the emperor could wear 'the true purple'. Extreme penalties were imposed on those not sanctioned to own purple garments, including the loss of property, title and even life.

It is believed that *tekhelet*, the blue dye used to colour Tallit prayer shawls came from a similar source and had the same method of production. However, the recipe was lost with the fall of Jerusalem in 70 CE.

The method of preparing Tyrian purple was also lost to Western civilisation after the fall of Constantinople during the Crusades of 1204. It wasn't rediscovered until 1998.

COLOUR + THE CLASSICAL WORLD

INDIGO

FOR MANY CENTURIES, INDIGO WAS THE
MOST IMPORTANT DYE IN THE WORLD.

It was used for textiles and wall paintings in the ancient world and it also helped support the British Empire. Heralded as Britain's most lucrative Asian export, indigo was used in vast quantities for mass-produced items like military uniforms.

Indigo is a molecule that is extracted from the leaves of the *Indigofera tinctoria* plant. First cultivated in the Indus Valley more than 5000 years ago the plant was called *nilah* (dark blue). The word indigo is derived from the Latin word *indicum* (coming from India).

The method of production of indigo dye has not changed in hundreds of years. Bundles of indigo leaves are put in a metre-deep vat that is filled with warm alkaline water. Logs are placed on top of the leaves to keep them submerged until the dye is drawn out through fermentation. When the water turns green, the liquid is transferred to a second tank. The indigo workers stand inside the tank and beat the liquid energetically with their legs to start the oxidisation process. A foamy blue froth covers the surface of the liquid as solid particles form and sink to the bottom of the vat. The precipitate is moved to a copper cauldron and boiled to reduce it to a paste. The dye is extracted, dried and sold as cakes of 'blue-gold'. Unlike most natural organic dyes, indigo can be used in pure, powdered form as a pigment rather than having to be made up as a lake.

Synthetic indigo was developed in 1880. By 1913 it had largely replaced the natural crop.

COLOUR + THE CLASSICAL WORLD

MALACHITE

THE NATURAL WORLD IS
FULL OF GREENS.

However, there were very few green pigments until the 19th century. Before that, there was malachite – a beautiful emerald green mineral found in copper mines. In Ancient Egypt, it was used as a cosmetic and as a colour to paint tombs.

Deposited in cavities by the slow percolation of water through copper-bearing rocks, malachite is found alongside its cousin, the blue mineral azurite. Both are basic copper carbonate and have almost identical compositions. In the Middle Ages, malachite was called *verde azzurro* (green-blue) in recognition of its close association with azurite.

Varying in colour from cold to bright yellow-green, malachite was used in Europe and the Far East for centuries. It has been found in medieval illuminated manuscripts dating back to the 8th century, and was a colour of great importance throughout the Renaissance.

To make malachite into a pigment, the mineral is carefully selected, crushed and ground to a powder. It is then washed in swirling water to separate the green particles, just like panning for gold. To be useful as a bright green, it must be ground coarsely. The more finely it is ground, the paler and more transparent the powder becomes. In *The Craftsman's Handbook*, Cennino Cennini advises that 'if you were to grind it too much, it would come out a dingy and ashy colour'.

By the end of the 18th century, malachite had been replaced with more easily obtained synthetic alternatives.

COLOUR + THE CLASSICAL WORLD

AZURITE

PURCHASERS OF AZURITE
HAD TO BE EXTREMELY WARY.

It is so similar to lapis lazuli – which was a much more expensive ore – that buyers could often be tricked. A careful apothecary would test a sample of the mineral by heating it until it was red-hot. Azurite turns black when it cools, while lapis lazuli does not.

Azurite is a bright blue mineral found in deposits of copper ore across northern and western Europe. Making a pigment from azurite is a relatively simple but laborious process. The hard stone has to be pounded in a mortar and pestle to a powder then washed with water and fish glue to remove impurities. Before being sieved, the powder must be washed multiple times in clean water.

When very finely ground, azurite is a pale sky-blue shifting towards green. A coarser grind gives a deeper blue colour, but the powder is very gritty. This makes the paint more translucent and difficult to apply. Many coats of coarse azurite are needed to achieve an opaque colour.

Until the arrival of the violet-rich blue of lapis lazuli from Afghanistan, azurite was the pre-eminent blue of European painting. Even afterwards, the incredibly high cost of lapis lazuli meant that azurite was still the most commonly used blue for masterworks of the Middle Ages and Renaissance.

The words azurite and azure are both derived via Arabic from the Persian *lazhward* (لاژورد), meaning 'blue'. In Europe, azurite was also known as lapis armenius or *citramarino* (a blue from this side of the sea), to distinguish it from *oltramarino* (a blue from beyond the seas) of the extract of lapis lazuli also known as ultramarine.

COLOUR + THE CLASSICAL WORLD

RED LEAD

THIS PIGMENT HAS BEEN USED FROM ROMAN TIMES THROUGH TO THE INDUSTRIAL ERA.

Red lead was used extensively in medieval manuscripts to paint small illuminations. In fact, the Latin word for red lead is *minium*, and *miniare* means 'to colour with minium'. Artists who painted with red lead were known as *miniator*, which is the source of our English word 'miniature'. The term 'minium' is now confined to the naturally occurring mineral that has the same composition as synthetic red lead. Minium deposits originally came from the Minius River in north-west Spain.

The Romans took great pleasure in its bold fiery hue, in particular for interior wall colouration. It is easily prepared by roasting white lead. At first it turns yellow, and then it becomes the deep orange-red of lead tetroxide.

Because red lead was cheap to produce, it was often used as a substitute for vermilion. Both mineral minium and synthetic red lead are highly poisonous, but the very opaque, bright, warm colour was invaluable as a primary red. It was the most commonly used red in medieval painting, until the introduction of synthetic vermilion. Red lead became almost obsolete after the introduction of the safer and more stable cadmium in the late 19th century. Until the end of the 20th century, red lead paint was used as an industrial anti-corrosion primer for structural iron and steel.

Red lead can only successfully be used as an oil colour. In fresco and other weakly bound paint films, it eventually reacts with atmospheric sulphides and turns black.

COLOUR + THE CLASSICAL WORLD

VERDIGRIS

THIS BLUE-GREEN PIGMENT IS
FORMED BY CORROSION.

Sheets of copper are suspended over a bath of vinegar, and the acetic acid fumes react with the metal to convert it into copper acetate.

Verdigris was the most vibrant green available to painters until the 19th century. Its English name is derived from *vert-de-Grèce* (Greek-green). The Germans call it *gruenspan* (Spanish green), the Ancient Greeks described it as 'copper flowers' and the Romans called it *aeruca*.

Verdigris was a popular but unstable pigment. It reacts dramatically with sulphur-containing pigments, such as ultramarine or orpiment, turning from green to dark brown. In Persia, miniaturist painters added saffron to verdigris to delay this darkening. Sulphur compounds in the air also corrupt verdigris, so it must be completely isolated with varnish.

In the 15th century, attempts were made to control this undesirable reaction by dissolving verdigris in hot oleoresins to make copper resinate. This new green was initially used with enthusiasm, but it fell out of favour as it became clear that in the presence of light and air it rapidly turned brown. Many paintings of the Renaissance have foliage and garments of a dull umber colour because the verdigris or copper resinate has corrupted over time. When viridian was introduced in the 19th century, the use of verdigris declined immediately. Because of its toxicity, it is now rarely sold.

COLOUR + THE CLASSICAL WORLD

CHRYSOCOLLA

THIS TURQUOISE-BLUE MINERAL WAS KNOWN
FROM ANCIENT EGYPT ONWARDS.

Chrysocolla was used in the classical world as a solder for gold and the Ancient Greeks named it after the words *chrysos* (gold) and *kolla* (glue).

Chrysocolla is composed of hydrated copper silicate. It is found in the same deposits as the other copper-bearing minerals, malachite and azurite. In its natural state, chrysocolla is similar to azurite but the colour is paler and slightly greener.

Due to its fragility, chrysocolla, unlike other coloured minerals, has limited use as a gemstone. It does, however, make a delicate, pale blue-green pigment that has been used throughout history. It has been found in wall paintings at Kizil in Turkistan and in Twelfth Dynasty Egyptian tombs. In oil painting it is very translucent, due to the pigment's low tinting strength. Chrysocolla was best employed in aqueous paints, such as egg tempera and especially watercolour, where, in the 16th century, it was called cedar green.

The pigment is prepared by systematically grinding the mineral from a coarse to fine powder. The powder is carefully washed and separated into different-sized particles by mixing it with large quantities of water. An egg yolk is often added to help with suspension. The coarser particles settle first, leaving the finer particles suspended. The process is repeated to separate the powder into different grades.

Skilful artisans could control the colour of the pigment by controlling the size of the particles. This allowed them to offer a wider range of colours from the same mineral.

IV.

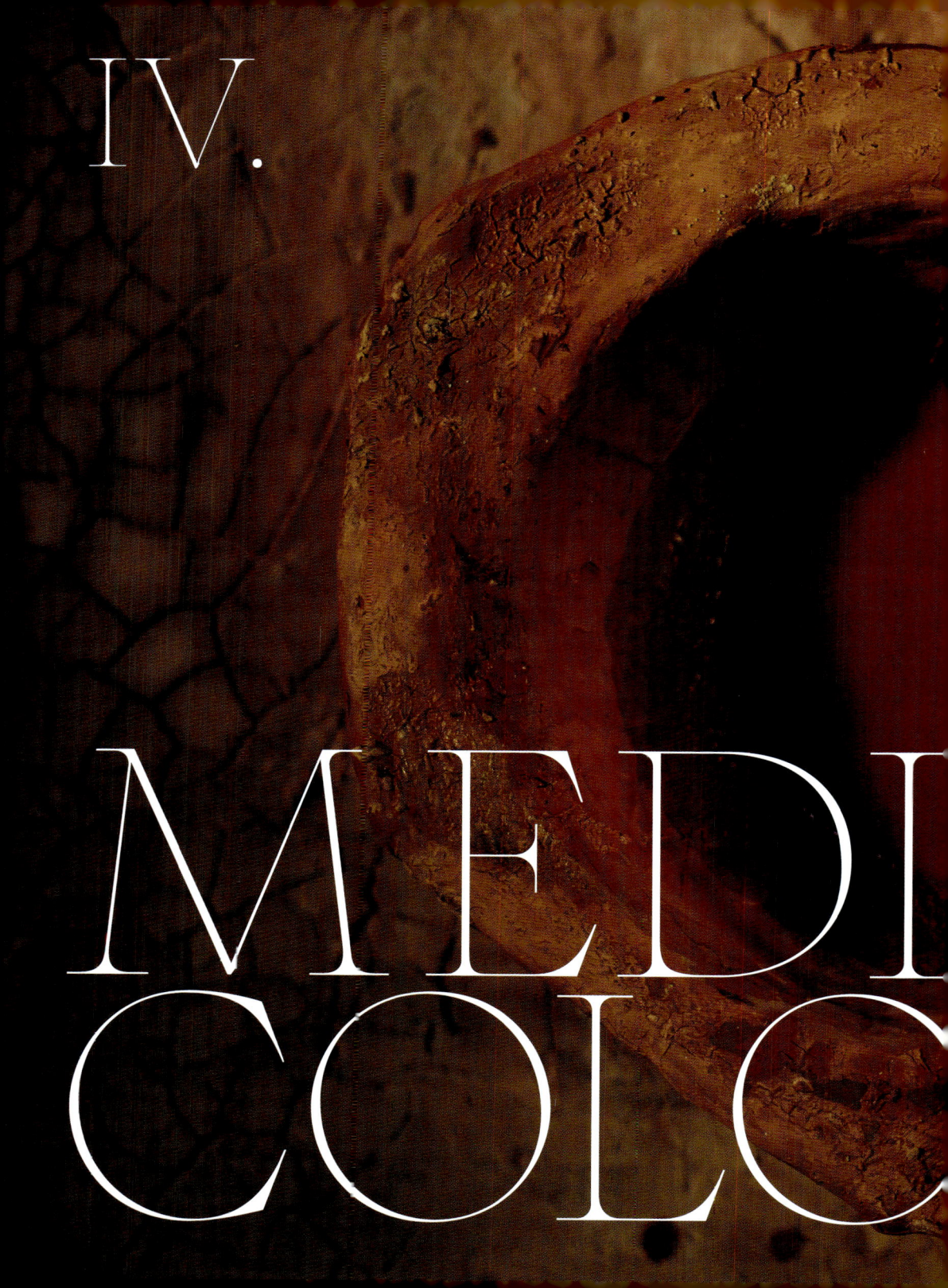

MEDI
COLO

EVAL
JRS

MEDIEVAL COLOURS

LAC

LIKE KERMES, LAC IS MADE
FROM A SCALE INSECT.

During the Middle Ages, this alternative to kermes arrived in Europe from India. Thousands of female *Laccifer lacca* insects infest fig trees, secreting a resin called sticklac, and encrusting the branches to protect their newly hatched larvae. The term 'lac' is derived from the Sanskrit word *Lākṣā* (the number 100,000), presumably due to the vast numbers of these insects.

The colour comes from the insect, not its resin. Lac is made by cutting the resin-laden branches, including the insects that are trapped in the sticklac. The insects are killed by exposure to the sun, and the sticklac is crushed, sieved and repeatedly washed in lye to release the red dye. The addition of a sulphate solution to the dye precipitates out as a lake pigment.

Lac was first imported into Europe around 1220. It quickly became the most important primary red lake for much of the Renaissance. It was used in such large quantities that the word 'lac' became a blanket term for all red dye-based pigments.

Also known as Indian lake, lac was a cheap and widely used pigment that, through careful alteration of the pH level, could be transformed into a range of reds, from orange through to violet hues. It is still used as a colourant for food and cosmetics, but its poor lightfastness has seen its use by artists disappear.

The lacquer called shellac is an insect-free form of the lac resin. If the insects are allowed to hatch and escape, the pale resin is refined into seedlac ('seed' refers to its pellet shape) and then further processed to shellac. It is used in varnishes, and as a coating for confectionery and pharmaceutical pills.

MEDIEVAL COLOURS

VINE
BLACK

THE MOST PRIZED CHARCOAL
IS MADE OF VINES.

Nearly any kind of wood will make charcoal but charring desiccated grapevines and stems makes vine black – the deepest rich blue-black there is.

The vines can be of any thickness but, because the wood shrinks as it turns into charcoal, thicker sprigs are normally selected. The bark is peeled away and if the vines are cut from fresh, living plants they are dried for a few days before charring.

The vines are packed tightly in tins, which are covered and sealed to prevent oxygen entering, and then roasted slowly over a fire. The vines cannot be burnt in the air, as this will turn them to ashes instead of carbon. It is important that the vines are charred thoroughly and if this isn't done, the colour will be brownish and the charcoal will have an unpleasant consistency. The conversion of vines to carbon takes several hours of roasting.

The finished charcoal can be used as a drawing implement or can be ground into a fine powder to make vine black pigment for ink or paint.

Charcoal was often used as a key component of cave painting with examples dating back at least 28,000 years. Since then, many cultures have used charcoal for art, camouflage and in rites of passage. Nowadays, charcoal (made mostly from willow) is used for drawing. Its softness is particularly useful for rapid mark-making in rich black.

MEDIEVAL COLOURS

KERMES

IS THIS A BERRY, A SEED OR
SOMETHING ELSE ALTOGETHER?

Kermes is, in fact, a wingless scale insect called *Kermes vermilio* that lives on the kermes oak tree of southern and eastern Europe. The insect is harvested by scraping it carefully from the branches of the oak tree. The red dye is extracted by crushing the resin-encrusted female insects and boiling them in lye.

Historically a very important crimson dye for textiles, kermes was imported from Mesopotamia by the Ancient Egyptians. Its trade routes covered the known world from Europe to China. When Spain was ruled by Ancient Rome, half of its taxes to the capital were paid in kermes.

The name comes from the Sanskrit word *krim-dja* (derived from a worm). In Hebrew it was called *tola'at shani* (worm scarlet). In medieval Europe, it was also called *baca* (berry). Alongside *granum* (grain) – another insect-derived red pigment of the Middle Ages – it shows how misunderstood the origin of this colour was.

Chaucer, the English poet known as the father of English literature, refers to a cloth that is 'dyed in grain', meaning dyed crimson by kermes or granum. Because of the strong, long-lasting nature of this colour, that phrase came to mean 'deeply or permanently dyed'. The word ingrained now means 'firmly fixed' and refers to attitudes and habits that are hard to shift. Kermes is also the linguistic root of the English words crimson and carmine, which both describe a deep purplish-red colour.

With the discovery of the New World in the 15th century, the stronger red of cochineal superseded kermes. By the 1870s, the use of European kermes as a colourant for textiles was gone.

MEDIEVAL COLOURS

DRAGON'S BLOOD

DRAGON'S BLOOD WAS CARRIED
IN SHIPS FROM YEMEN TO EGYPT.

Elephants have continual warre against Dragons, which desire their blood, because it is very cold: and therefore the Dragon lying awaite as the Elephant passeth by, windeth his taile, being of exceeding length, about the hinder legs of the Elephant, and when the Elephant waxeth faint, he falleth down on the serpent, being now full of blood, and with the poise of his body breaketh him: so that his owne blood with the blood of the Elephant runneth out of him mingled together, which being colde, is congealed into that substance which the Apothecaries call Sanguis draconis, *that is Dragons blood.'* [2]

This fantastical tale of the origins of dragon's blood is recounted by the 16th-century navigator, Richard Eden. In reality, dragon's blood was a garnet-red resin exuded from *Dracaena cinnabari*, the Socotra dragon tree.

The Ancient Roman author Pliny is the source of this colour's fanciful name. He refers to it in his book *Natural History*, written in 77 CE – just two years before his death in the eruption of Mount Vesuvius.

Dragon's blood has been used as a colourant, medicine and in alchemy. In medicine it was employed for curing diarrhoea, skin disorders and fevers. As a colour it was mainly used to tint varnishes. It was applied over gold to create a ruddier effect, was used by Stradivarius for tinting his violin varnishes and by furniture makers for the colouring of lacquers.

MEDIEVAL COLOURS

LAPIS LAZULI

HOW COULD A COLOUR COST
MORE THAN GOLD ITSELF?

Ultramarine is extracted from the natural rock, lapis lazuli (blue stone) and was sourced almost exclusively for 6000 years from Afghanistan. The name comes from the Latin word *oltramarino* (a blue from beyond the seas).

Lapis lazuli's colour comes from the blue mineral, lazurite. Cennino Cennini describes it as 'a colour illustrious, beautiful, and most perfect, beyond all other colours'. Ground lapis lazuli results in a fairly weak blue (unless the stone is pure lazurite). Despite this, it appears from the 6th century in Byzantine manuscripts and in Afghan wall paintings.

Iron pyrites and white calcite impurities must be removed from lapis lazuli to provide the exquisite ultramarine that was so beloved by Renaissance painters. The extraction process is believed to come from a 9th-century Arabic alchemical source. This complicated and time-consuming recipe describes how to grind lapis lazuli and mix it into a paste of wax and resins. The paste is laboriously kneaded in an alkali solution to draw out the blue lazurite. Using this process, high-quality ultramarine appeared in Western art at the beginning of the 13th century.

For every 100 grams of lapis lazuli, only four grams of genuine ultramarine pigment is collected. Given its incredible cost, the colour was reserved for only the most important subjects, in particular, depictions of the Virgin Mary. Most painters used it in thin glazes over an opaque underpainting to reduce costs.

With the invention of cheap, synthetic ultramarine in the 19th century, its natural counterpart rapidly fell out of use.

MEDIEVAL COLOURS

PEACH BLACK

THIS CARBON BLACK IS MADE
BY CHARRING PEACH STONES.

Used from the Middle Ages onwards, this finely ground black pigment is one of a family of impure carbon blacks that are derived from the shells or the kernel of fruits like cherry, almond, walnut and coconut.

Peach black is close in chemical composition to other char blacks, such as the charcoals of vines and willow. Unlike lamp blacks, which are made by creating airborne soots, char blacks retain some of the original structure of their raw materials.

The manufacturing process of peach black is also very similar to other char blacks. When the peach stones are roasted it is essential that air is excluded. If air gets in, the stone will burn and turn into white ash instead of carbon.

Opaque and with excellent covering power, this deep blue-black pigment was recommended for watercolours. Its very slow drying times – a result of residual tarry material in the pigment – meant it was rarely used in oil painting. Due to the expense of collecting the stones and processing the pigment, peach black has generally fallen out of production.

Peach and other fruit-stone blacks did, however, find a life-saving application in World War I. When Germany started using deadly chlorine gas against the Allied troops in the trenches, an American chemist found a solution by filling gas masks with activated charcoal made from natural fibres such as those found in peach stones. The Red Cross organised the collection of millions of peach stones that were turned into charcoal, and consequently saved countless lives.

MEDIEVAL COLOURS

LEAD TIN YELLOW

IF LAPIS LAZULI WAS THE KING OF THE RENAISSANCE PALETTE, THEN LEAD TIN YELLOW WAS ITS QUEEN.

For 400 years it was the most important yellow for artists. Produced by heating a mixture of lead oxide and tin oxide at about 800°C, lead tin yellow is made in two different forms. The earlier form (confusingly known as Type II) contains silicon and was initially used in the production of coloured leaded glass in the Middle Ages. The later form (Type I) was most frequently used in paintings, with the earliest examples being from Florence at the beginning of the 14th century.

Although often used alone, lead tin yellow was also added to green and earth pigments to create grass and foliage colours. Care must be taken not to mix it with sulphide-containing pigments, as this turns the yellow to black.

Lead tin yellow is a modern term. During the Renaissance it was known in Italy as *giallolino* (from the Italian for yellow, *giallo*). In English it was called 'general' and in the north of Europe it was called 'massicot'. The historical habit of associating all yellow pigments made from lead oxide created confusion for future researchers. The various names for lead tin yellow later became synonymous with pigments made from lead antimonite (like Naples yellow) or lead oxide.

Inexplicably, lead tin yellow fell out of favour in the mid-18th century, and even the knowledge of its presence in painting was lost. It was rediscovered in 1941 by the German scientist, Richard Jacobi. With the help of sophisticated testing equipment, he was able to precisely identify the chemical constituents of lead tin yellow and restore this pigment to the history of colour.

MEDIEVAL COLOURS

VERMILION

THIS RED WAS KNOW TO TURN
MINERS INSANE.

In 1566 the King of Spain sent condemned criminals to serve their sentences at the mercury mine of Almaden. The dangerous working conditions at the mine and the well-known nature of mercury poisoning had made it difficult to find willing labourers. In the second half of the 16th century, one-quarter of the prisoners died before being released.

The prisoners were mining the mineral cinnabar, from which mercury is extracted. Cinnabar is the name given to both naturally occurring mercuric sulphide and the opaque bright red colour that it gives rise to. Vermilion is its synthesised form.

The recipe for making vermilion was introduced to Europe by an Arabic alchemist around the 8th century. Several centuries later, Cennino Cennini described the process:

Take one part of mercury and one of sulphur. Put it in a glass bottle, thoroughly clad with clay. Put it on a moderate fire and cover the mouth of a bottle with a tile. Close it when you see yellow smoke coming out of the bottle, until you see the red and almost vermilion-coloured smoke. Then take it from the fire and the vermilion will be ready. [3]

The resulting lump is black, but when ground with water on a slab a fiery red colour develops. Vermilion's marriage of two fundamental substances – sulphur (which was believed to be base gold) and mercury (quicksilver) – made it incredibly interesting to alchemists.

MEDIEVAL COLOURS

SMALT

THE ORIGINS OF SMALT LIE IN THE
ANCIENT COLOUR EGYPTIAN BLUE.

The name comes from the Italian term *smaltare* (to melt). Like Egyptian blue, smalt is produced by heating lime, silica and potash, but smalt's colour comes from the replacement of cobalt for the copper in the formulation.

Heated together till molten, the liquid glass is poured into cold water, where it shatters into small pieces. The glass is sieved and ground to yield different-sized particles of varying colour. To retain a strong, deep blue colour, the powder must not be ground too fine. However, a coarse grind makes the resulting pigment gritty, and it is a difficult material to paint with. It is also a relatively transparent pigment; so multiple layers have to be applied to a surface to create an opaque colour.

Smalt was in use in Europe from at least the 15th century. For many years it was believed to be a European invention, but recent research has uncovered much earlier applications. Cobalt has been found in the colouring of glass from the ancient world of Mesopotamia and Egypt where it was introduced alongside the development of vitreous enamels for jewellery and ceramics.

In Europe, smalt was popular because of its low cost. It was used as a substitute for the more expensive ultramarine and azurite, but unfortunately its blue colour diminished when it was mixed with oil. The yellowing of the binder reduced it to grey.

Smalt's popularity ended in the 19th century. With the introduction of modern cobalt blue and synthetic ultramarine, artists had access to more powerful and easily applied blue pigments.

MEDIEVAL COLOURS

SAFFRON

FOR A MEDIEVAL MANUSCRIPT ILLUMINATOR, THE MOST IMPORTANT YELLOW CAME FROM THE STAMENS OF THE CROCUS FLOWER.

It took 8000 handpicked flowers to collect just 100 grams of red saffron threads. Saffron produced a strong, pure, translucent yellow that could be used to imitate gold leaf.

Originally known as Persian yellow, saffron was used by the Ancient Sumerians as a perfume and medicine. The Ancient Egyptians used it for dyeing mummy bandages and Roman emperors used it to perfume their baths. It has been used since antiquity as a dye for fabric, colouring the robes of Chinese emperors, and in wine, food and cosmetics. Saffron was also reputed to be the colour of love.

Aside from the difficulty of collecting enough stamens, most recipes for saffron paint are extremely simple. The saffron strands are soaked overnight in glair and allowed to infuse. It was also commonly mixed with blues to produce a wide range of green tones. Cennino Cennini says that a blend of saffron and verdigris produces 'the most perfect grass colour imaginable'. However, because saffron fades, the medieval manuscripts we are now left with feature blue trees, grass and clothes where green was intended.

Saffron's use as a pigment effectively ended with the advent of cheaper lightfast synthetics. Now it is best known as a flavouring and colourant of foods.

MEDIEVAL COLOURS

BLUE VERDITER

A MEDIEVAL ARTISAN WHO WANTED
A BRILLIANT BLUE HAD TO HAVE
DEEP POCKETS.

The artists had ultramarine and azurite, but these were too expensive for decorative work or anything but the most highly paid commissions.

There was, however, another option. Blue verditer, a synthetic azure blue pigment, is prepared by the reaction of a copper nitrate or copper sulphate with calcium carbonate and sal ammoniac. This pigment was probably more significant in medieval painting than all the other blues put together. It is certain that, for every kilo of ultramarine used in the Middle Ages, many tonnes of blue verditer were used, often under the name 'refiner's verditer'.

This basic copper carbonate pigment was first made in the 15th century and was in constant use for panel painting, distemper and oil-based interior house paints until the 19th century. Its origins are in silver mining. A copper nitrate solution (a by-product of the process of separating silver from copper) was accidentally spilt onto chalk, which immediately turned green. This observation lead to a controlled process for manufacturing the original green verditer. The English manufacturers later perfected the process to make blue verditer.

Blue verditer is chemically identical to ground azurite, but its particles are more rounded and regular in size. Unfortunately, they share the tendency to turn green in the presence of even weak acids.

Also known as blue bice and blue ashes, blue verditer has now been superseded by more stable synthetic blues. This once-famous colour is now almost unknown.

MEDIEVAL COLOURS

GRAPHITE

WITH ITS DARK-GREY COLOUR, GRAPHITE
USED TO BE CONFUSED WITH LEAD.

In fact, until 1789 it was called blacklead or plumbago (from *plumbum*, Latin for 'lead') and only a few decades ago, children began their handwriting journey with 'lead' pencils. The modern term comes from the Greek word γράφω (*graphos*, or 'I write'). Like diamond, graphite is actually a crystalline modification of carbon, but its soft nature makes it ideal for mark-making.

Until the mid-16th century, because of its limited availability, graphite was seldom used for drawing or writing. In the classical world, genuine lead was used instead and in the Renaissance silverpoint was used.

Around 1565, an enormous deposit of pure graphite was discovered in Borrowdale, England. Even today, the local sheep's coats are grey from rubbing against graphite-rich walls. During the Elizabethan era, this graphite was used to line moulds for cannonballs. This resulted in rounder, smoother balls that could be fired farther and contributed to the strength of the English navy. Because of its importance, the graphite mine and its production were strictly controlled by the Crown.

At this time, graphite's popularity for drawing increased but, because it is so soft, it needs a casing to stop it breaking or leaving marks on the artist's hands. Initially, sticks of graphite were wrapped in string or sheepskin for stability but the first wood-encased pencil was invented soon after when two halves of timber were glued together around a graphite core.

In 1795 the French painter and balloonist, Nicolas-Jacques Conté, broke the British monopoly on solid graphite by inventing the modern pencil.

MEDIEVAL COLOURS

NAPLES YELLOW

NAPLES YELLOW IS SAID TO
ORIGINATE FROM MOUNT VESUVIUS.

Other yellowish minerals did occur on Mount Vesuvius, so at first glance this seems to be a plausible claim. Unfortunately, the facts get in the way of a good story as there has never been any evidence of its presence there.

First created by the Ancient Egyptians, the pigment we now know as Naples yellow was confined to application on vitreous enamels rather than as a pigment for painting. It is made by heating lead and antimony compounds together to produce a range of yellows from lemon to a dusty yellow-orange. Lead-antimonite yellow has been identified in objects and paintings from Egyptian, Mesopotamian, Babylonian, Greek and Roman cultures.

In European art, Naples yellow is estimated to have been in use for easel painting since at least 1600, although there is some uncertainty about this because of the fluidity of pigment names throughout history. The confusion of terms for lead-oxide yellows has resulted in the incorrect identification of the pigment in paintings over the centuries. Sometimes the Italian term *giallolino* (used in the Middle Ages) refers to a yellow made from lead and tin, and sometimes it refers to Naples yellow. Naples yellow was most frequently used during the period 1750–1850. After this, it was gradually replaced by chromes and cadmiums.

Genuine Naples yellow continues to be manufactured in limited quantities. However, the name is now used more loosely by paint manufacturers to indicate a colour made by mixing together other pigments, most frequently cadmium yellow, zinc white and red ochre.

V. WRITING INKS

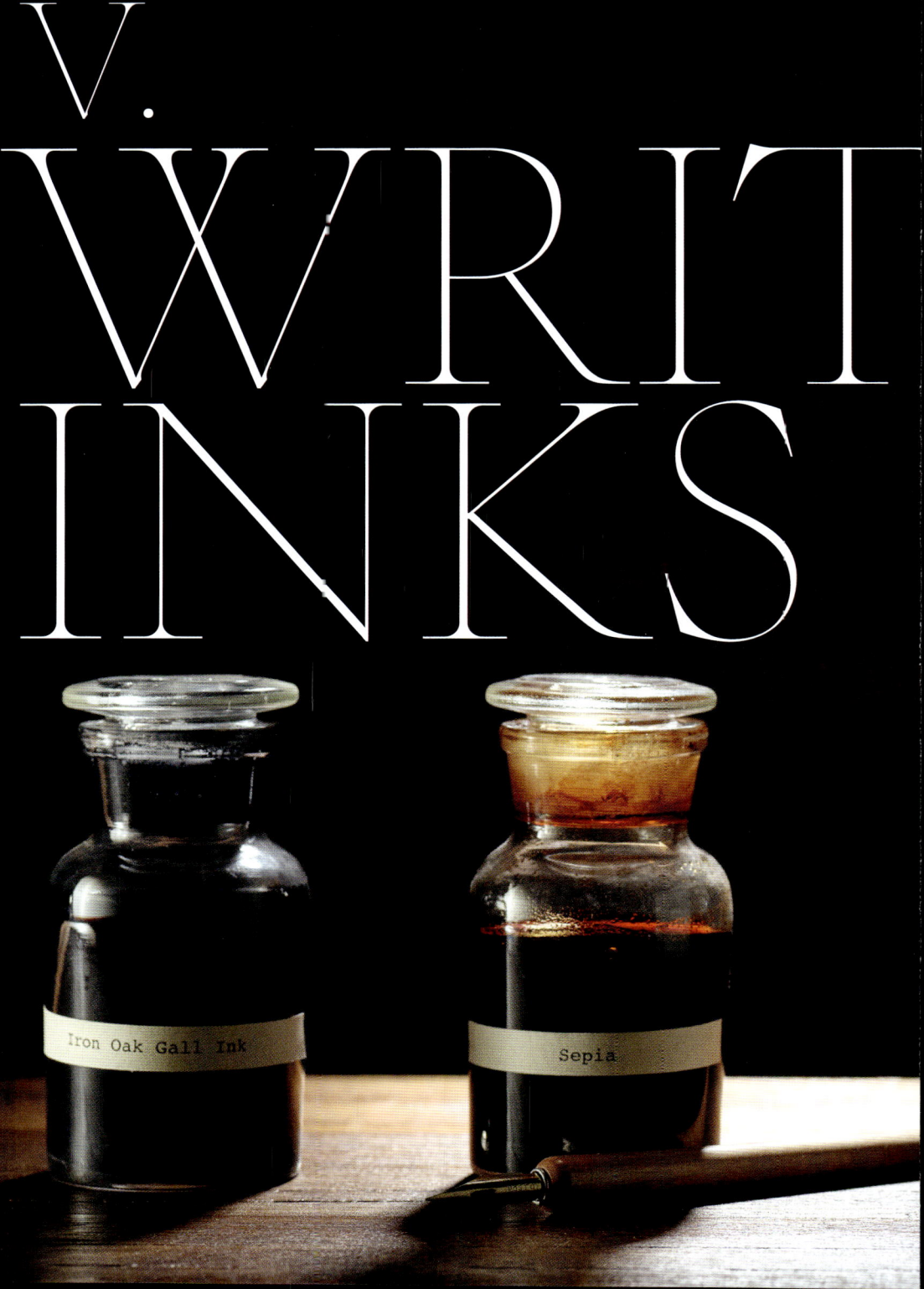

ING

Walnut Ink   Bistre

WRITING INKS

GALL INK

THIS MEDIEVAL INK
STARTS WITH A WASP.

In spring, the gall wasp punctures the soft young buds of the oak tree and lays her eggs. The tree forms small nut-like growths around the wasp holes. It is these protective oak galls that form the basis of this intense black ink.

Gall ink was the standard writing and drawing ink in Europe from at least the 5th century and remained in use well into the 20th century. The process for making it is likely an adaptation of textile dyeing recipes that go back to antiquity.

The oak galls are harvested from the dyer's oak tree and are graded according to their maturity and tannic content. Blue and green grades are younger, contain the wasp larvae and have a high gallotannic acid content. Only in the white gall has the mature insect chewed a hole through the gall in order to escape.

The ink is prepared by fermenting the crushed oak galls in water to release the concentrated brown gallotannic acid. The addition of green copperas (iron sulphate) darkens the ink and helps with permanency. After filtering, a binding agent of gum arabic is added to the ink to control how it flows from a nib or brush. A well-prepared ink gradually darkens to an intense purplish black. The resulting marks adhere firmly to the parchment or vellum and – unlike Indian ink or other formulas – can't be erased by rubbing or washing.

In the UK, gall ink is still used for all official certificates of birth, marriage and death.

WRITING INKS

BISTRE

THIS INK COMES FROM THE
SOOT OF BURNT BEECHWOOD.

Bistre is a dark brown, transparent pigment that is made from a mixture of flame carbon, charred wood and uncarbonised tarry material. It was used from the 14th century in manuscript illuminations, and in the 17th and 18th centuries for wash drawings and watercolour paintings.

The colour of the ink is determined by how much resin is present in the beechwood. After being turned into charcoal, the soot is mixed with water and cooked to thicken. It is important not to boil the mixture while it is being reduced, as the natural colour of the resin may be destroyed. The resin separates from the liquid and is cleaned. Multiple washes are needed to filter away the dirty excess soot. Once the mixture is clean and rich in resin, it is reduced again until it reaches the desired consistency.

There is little mention of the term bistre before the 17th century. Prior to this, it was variously known as soot brown, *fuligno* or *caligo*. Caligo is Latin for 'darkness', from the Ancient Greek κελαινός (*kelainós*, 'dark, black'), which in turn is derived from the Sanskrit word *kāla* (black).

Bistre was used primarily in water-based paints and drawing inks because asphaltum – a much more powerful dark brown pigment – cannot be added to water due to its oiliness. As a result of its organic material content, bistre is not very lightfast. It was replaced by more lightfast pigments in the 19th century.

WRITING INKS

SEPIA

SEPIA COMES FROM THE
INK SACS OF CUTTLEFISH.

When the Sepia is frightened and in terror, it produces this blackness and muddiness in the water, as it were a shield held in front of the body. [4]

What Aristotle is describing here is the defence mechanism of the cuttlefish. When attacked, the cuttlefish injects an inky cloud into the water, much like a smoke screen, to help it make a rapid escape from predators. The dark colour comes from its main constituent, melanin.

This natural ink can also be made into a deep warm brown pigment named after *sēpía*, the Greek word for cuttlefish. Used as a writing ink from Ancient Roman times onwards, sepia remained in common use until the 19th century.

The ink sacs of the cuttlefish are carefully extracted from the bodies and dried to prevent putrefaction. The solidified sacs are then dissolved in a weak alkali solution of potash and the resulting solution is filtered. The pigment is precipitated with dilute hydrochloric acid, washed, filtered again and dried.

Sepia only become popular as a drawing ink from the Renaissance onward. It was first introduced as a watercolour around 1780 and generally replaced bistre as a medium for making wash drawings.

The drawing ink now sold under the name sepia replicates the original colour but is made with more easily produced modern pigments. Genuine cuttlefish ink is still used to add flavour and colour to foods.

WRITING INKS

WALNUT

THIS INK IS MADE FROM
THE HUSKS OF WALNUTS.

Used since the Middle Ages, walnut's rich, velvety brown colour comes from the naturally present dyes and tannins.

Although the immature green husks can also be used, the walnuts are traditionally collected when they are mature as the older, shrivelled black husks are softer and easier to remove from the shell.

There are two methods for obtaining the deep brown ink from the husks: hot or cold extraction. In the hot process, the husks are simply covered with water, brought to the boil and simmered for a few hours to extract the colour. The solution is filtered and simmered again to reduce the liquid and the colour is concentrated. A precipitate containing the darker portions of the colour often develops. This solid can clog writing instruments, especially fountain pens. However, without it, the ink is paler than the colour created by the cold process.

The cold process involves fermentation. A nonferrous container is filled with the husks, which are completely covered with distilled water. A lid is placed on top and the husks are left to ferment for at least two months. After this time, the resulting black liquid is filtered to remove the husk solids and evaporated until the desired colour is achieved. This can take many weeks, so very gentle warming is often employed to speed up the process.

Walnut ink permanently stains skin and has been used throughout history as a colourfast and lightfast dye for textiles.

VI.
DYES,
LAKES
PINKS

S+S

DYES, LAKES + PINKES

ARZICA

MADE FROM WELD, THIS
PIGMENT IS CALLED ARZICA.

Weld, also known as dyer's rocket, is the oldest European dye plant and was commercially cultivated well into the 20th century. It is particularly valued for the clear and intense yellow it produces when dyeing silk.

Arzica is the lake pigment made from the weld dye. The whole plant – flowers, stems and all – is dried, broken up and stewed in a weak solution of alum to extract the colour. A solution of potash is added and the insoluble lake pigment is formed. This is allowed to settle, then it is washed, filtered and dried.

Because arzica is translucent, it is an ideal pigment for glazing. Medieval colour-makers held weld lakes in high esteem, but they were often used unobtrusively. They were mixed with a range of blues to give greens and added to duller yellow pigments, such as yellow ochre, to add more chroma.

Sometimes in the making of arzica chalk, eggshells or white lead were included to add opacity. This bright, solid yellow provided a good (and harmless) substitute for the poisonous orpiment, which contains arsenic. In fact, the name arzica is most likely a linguistic corruption of *arsenikon*. The use of the same name for different pigments of a similar colour was a common practice.

Unfortunately, arzica, like most vegetable-based lakes, is fugitive and fades under direct sunlight. It is no longer used as a pigment, as it has been superseded by more lightfast yellows.

DYES, LAKES + PINKES

BRAZILWOOD

HOW DOES A COUNTRY COME
TO BE NAMED AFTER A WOOD?

Throughout the Middle Ages, artists who wanted cool reds looked to lake pigments made from brazilwood, a hardwood tree that provides a deep red dye.

For centuries, the isle of Serendip (now known as Sri Lanka) was the main source of brazilwood, but after the discovery of the New World, vast quantities were shipped back to Europe from South America. The Portuguese even named their colony Brazil after this valuable commodity.

The word brazil comes from the same root as the French word *brazier* and refers to the glowing red colour of the dye.

Brazil was sold in blocks and turned into a fine powder by scraping it with a piece of glass. This painstaking process was followed by steeping in water and the addition of alums, metal muriates and alkalis. Hundreds of variations on extraction methods create a wide range of shades. Adding more alum leads to warmer, more orange colours. Increasing the lye content creates cool crimsons. Adding chalks, crushed eggshells or white lead results in opaque pinks.

It cannot be underestimated how much brazilwood dye and pigment was used in the Middle Ages. It was a major source of reds. Brazil was as important as kermes and granum, but it was much more common because it was cheaper and easier to use.

However, it was not particularly lightfast and, like kermes and lac, it was largely replaced after the discovery of the more brilliant cochineal and the more permanent madder.

DYES, LAKES + PINKES

LOGWOOD

PURITANS WERE THE FIRST TO
WEAR FASHIONABLE BLACK.

In the 16th and 17th centuries, the rise of Puritanism led to the vogue for garments that were devoid of colour. To be truly pious, one had to wear black. The problem was that there were no true black dyes. That all changed when the Spanish brought logwood back from the New World.

As a genuine black fabric dye, logwood was of great economic importance from the 17th century onwards. Hundreds of tonnes of logwood were exported and the ships carrying it back to Spain were a constant target for privateers. This activity was encouraged by the English Crown, as Spain and England were at war throughout this period. With the peace treaty of 1667, the Spanish Crown granted trading rights in return for the British suppression of piracy.

The privateers, in need of income, turned to logging. The colony of British Honduras, now known as Belize, grew from English logging camps that were established to export logwood. The best trees were the old ones, because they had less sap and were easier to cut. As the young adventurer, William Dampier explained:

The sap is white and the heart red. The heart is used much for dyeing; therefore, we chip off all the white sap till we come to the heart ... After it has been chip'd a little while, it turns black, and if it lyes in the water it dyes it like ink. [5]

With the introduction of synthetic dyes in the 19th century, logwood's central role in dyeing disappeared.

DYES, LAKES + PINKES

STIL DE GRAIN

STIL DE GRAIN WAS MADE FROM THE
BERRIES OF THE BUCKTHORN BUSH.

Like all lake pigments, this dye is extracted with an alum solution and converted into a pigment by adding a solution of potash.

Colour-makers in the Middle Ages and later became very skilled at producing a wide range of colours from this one berry. The final colour can be controlled by the time of harvesting, the addition of tin, copper or iron salts, and the temperature of the dye solution. When the solution is held at 50°C, a lemon-yellow lake is created. Kept at 100°C, it produces an orange colour. Genuine sap green can also be made from unripe buckthorn berries.

In the Middle Ages, stil de grain was sold as a dense syrup in bladder sacks instead of being dried and sold as powder. During the 18th century it was used extensively in France and England as an oil colour.

Stil de grain is a fugitive pigment, which means it is not permanent. However, stil de grain used in medieval manuscripts has been protected from light and moisture, safely enclosed in the pages of a book.

Stil de grain was also known as yellow madder, Dutch pink, brown pink and English pink. Stil de grain is now obsolete, having been replaced by modern lightfast colours.

DYES, LAKES + PINKES

MADDER LAKE

THIS PIGMENT WAS INTRODUCED
TO ITALY BY THE CRUSADERS
RETURNING FROM THE HOLY LANDS.

Cultivated across Europe from at least the 13th century, the madder plant was used to create a permanent red textile dye. Four centuries later, a beautiful new red pigment derived from the dye was introduced: madder lake.

The process of creating this exquisite red is long and complicated. Madder lake is made by fermenting or boiling the roots of the madder plant (*Rubia tinctorum*) in an alkaline solution to extract the main colouring agents. The dye takes at least three years to properly develop within the root of the plant. The mature plant is dug up and the older roots are selected, cut off and dried. The roots can be over a metre long and up to 12 mm thick. The plant is replanted to allow it to continue growing.

The alchemy of drawing out and converting the dye into a pigment is complex. Depending on chemical selection, temperature or how long it ferments, a wide range of pinks and reds through to crimsons can be achieved. The refinement of the production methods reached its height in 19th-century England through the work of dye-maker George Field

The most famous pigments are alizarin crimson and rose madder. The extraction of alizarin creates a deep crimson colour that is useful for skin tones. Its highly translucent nature also makes it excellent for glazing. Extracting pseudopurpurin creates the delicate pink of rose madder.

In 1869, German chemists Carl Graebe and Carl Liebermann synthesised alizarin. This effectively put an end to the cultivation of madder for commercial dyeing. Madder is still grown in limited quantities for dyeing traditional Persian rugs and textiles.

DYES, LAKES + PINKES

COCHINEAL

IS THIS RED REALLY
MADE OF BLOOD?

Used in the Americas for dyeing textiles as early as 700 BCE, this vivid scarlet was treasured by the Incas and the Aztecs. One of the reddest dyes that the natural world has ever produced, the crimson dye is carminic acid, which is produced by female cochineal to deter other insect predators. Around 14,000 insects are needed to make just 100 grams of carmine lake pigment.

When the Spanish conquistador Hernán Cortés entered the Aztec capital of Tenochtitlán in 1519, he found its markets full of bales of delicate yarns dyed a sensational crimson red. It turns out the Aztecs afforded enormous prestige to cochineal and tributes paid to the emperor Montezuma by each state were bags containing millions of dried cochineal. After the Spanish conquest of the Aztec Empire, cochineal also became a very important export for the Spanish. They protected their exclusive supply by disguising the red dye's origins in mystery, spreading the story that cochineal was a pea-like vegetable. This is more plausible than it might sound, as the dried insects look very much like shrivelled berries. This deliberate misinformation made cochineal production one of the best-kept trade secrets of all time. It became the third-greatest traded product from the New World, after gold and silver. Even today, the only crop in Latin America that can compete with cochineal in price is cocaine.

After synthetic dyes were invented in the late 19th century, cochineal production almost vanished. However, in recent times, due to health fears over artificial food additives, a renewed interest has lead to a return in its popularity. It is now used in sweets, fruit juices, cosmetics and liquors like Campari.

THE FARMERS OF COCHINEAL

Fig. 1. Indio que recoge la Cochinilla con una colita de Venado, *Fig. 2.* dicha *Fig. 3.* Xicalpestle en que aparan la Cochinilla.

DYES, LAKES + PINKES

The legend of cochineal's creation comes from the land of the Cloud Men – modern-day Oaxaca in Mexico. The tale unfolds before the time of humans, when gods inhabited the earth. Two powerful deities fought to the death over a valuable crop of cacti and during the fatal battle, their blood was splattered across the cactus fields. Their brothers and sisters descended to carry the bodies back to heaven on a bed of clouds, but their spilt blood became the tiny cochineal insects that produced the most sought after natural dye in human history. The Aztecs called it *nocheztli*, from the Nauhatl words *nōchtli* (cactus) and *eztli* (blood).

In Oaxaca, cochineal is still a major commercial concern, produced by small farms that harvest cochineal for local textile dyeing as well as for limited export to the rest of the world.

The main host plant for the cochineal insect is the nopal cactus, also known as the prickly pear. As early as the 12th century, the Aztecs domesticated the insect and, through selective breeding, increased the strength of the dye they produced. Plots were prepared with fertilisers, plantations were protected from wind, rain and frost, and predators such as spiders, rats and lizards were removed.

Modern cochineal farming is relatively unchanged from the days of the Aztecs, except that the process is now more controlled. Today, commercially farmed plants are grown in barns to control access to water and regulate their temperature. These cochineal-laden cacti are tended carefully as the young insects grow, covered in a white gossamer of nymph wax. Through cultivation, nopal cacti have become spine-free, making harvesting easier.

In the wild, the cochineal are carried from plant to plant by the wind, whereas the commercial process involves controlled reseeding. This can be done in a couple of different ways: some farmers gently transfer the cochineal with fox-hair brushes and put them onto healthy plants, while others place small baskets, called 'Zapotec nests', full of fertile females onto the cacti. Once safely on the nopus cactus, the females leave the nests and await fertilisation by the males. Females have a round body and look similar to woodlice (cochinilla means 'woodlouse' in Spanish), whereas males are thin, flying insects that only live for a few days – their sole purpose to fertilise the females.

THE FARMERS OF COCHINEAL

107

The whole cycle of growth from birth to harvesting takes three months. During commercial cultivation, the cochineal-infected cacti are kept at a constant temperature of 27°C, as the insects are very susceptible to cold and extreme heat. Once they are fully grown, the insects are harvested. Traditionally they were spooned from the cactus, but modern methods use compressed air to remove them. The cochineal are then laid out in the sun for a week to dry, then packed for distribution or sold to local weavers. The local Oaxacan dyers crush the cochineal into a powder using a pestle and mortar to extract the dye more easily. Traditional dyeing uses natural components to produce different dye shades – adding water produces a dark crimson colour, whilst adding lime juice creates exquisite scarlets.

When the Spanish conquered modern-day Mexico in 1519 they realised that cochineal's concentrated colour was far superior to the European red dyes of kermes and grain. Within just a few decades, cochineal made a significant impact across Europe. Its arrival coincided with the disintegration of the old dye guilds, and the movement of highly skilled people around Europe helped spread the use of the dye across the continent and beyond.

Cochineal's global importance continued for 300 years until the mid-19th century, when – calamitously for the cochineal farmers – synthetic dyes were invented. Cheaper than their natural counterparts, they quickly dominated the market and cochineal farming almost completely disappeared. Only when health issues with synthetic dyes used in foods and cosmetics were identified in the 1980s did cochineal make a comeback.

However, it wasn't the Oaxacans who benefited from this resurgence. Other Latin and South American countries, Peru in particular, created private commercial companies that took advantage of agricultural science to maximise production. They now dominate world trade. The Oaxacans continue to rely on the traditional system of farmers' collectives using small household plots. Although romantic, it cannot compete with modern technology and large businesses.

Whether the Oaxacan farmers can enjoy some of the future rewards of commercial supply is yet to be seen, but these farmers deserve, at the very least, to be recognised as the cultural keepers of an ancient tradition.

COCCINELLA, AND COCCUS.

Fig. 1. to 15. different Species of Coccinella. Fig. 16. to 21. male and female Coccus. Fig. 22. to 27. the Insect supposed to feed on the Coccus.

MYSTI
COLO

VII.

CURIOUS
OURS

MYSTERIOUS COLOURS

INDIAN YELLOW

FOR YEARS, THE INGREDIENTS OF
INDIAN YELLOW WERE A MYSTERY.

Speculation abounded amongst colour-makers and artists. Some noted the smell and believed that it was made from the urine of camels or snakes. Known in India since the 15th century, Indian yellow had many names. These included *peori*, *puree* and, most interestingly, *gogili*, which is a corruption of a Persian term meaning 'cow earth' and perhaps offers a clue about where it was first invented.

It wasn't until 1883 that the rumours were supposedly settled. Indian yellow was made exclusively in the Indian village of Mirzapur. A Bengali civil servant, T.N. Mukharji, travelled there to investigate and the letter he sent back to the British authorities detailed the extraordinary process. Indian yellow was prepared from the urine of cows that had been fed only on mango leaves. The urine was collected in small earthen pots and then concentrated over a fire. The liquid was strained and the sediment was collected and formed by hand into balls before being dried in the sun.

Given no food apart from the mango leaves the cows were in a very poor state of health. This disgusted dairy farmers, who called the colour producers 'cow destroyers'. Curiously, successive attempts over the years to confirm this story have proved inconclusive.

A clear and luminescent but scarce pigment, Indian yellow was used by European painters up until the 1900s. The introduction of lightfast modern pigments in the early 20th century lead to its rapid decline and eventual disappearance.

MYSTERIOUS COLOURS

GAMBOGE

THIS IS A POISONOUS GOLDEN PIGMENT.

Gamboge is derived from trees native to South-East Asia. Using a method similar to rubber extraction, a deep incision is made in the trunk of the *Garcinia* tree. A hollowed-out bamboo tube is carefully placed beneath the cut and a milky, yellow resin fills the bamboo mould. The resin is roasted over a fire to evaporate the moisture and once dried, a solidified resin cylinder is removed. This is then ground down to produce a bright yellow powder.

The resin takes its name from Camboja, the old form of Cambodia, which was the principal country of supply. It was used in water-based inks from the 8th century in Japan, China and Thailand. At the beginning of the 17th century it was imported into Europe where artists took it up as a transparent warm yellow. The Flemish painters used it in oil colours, but it was found to work best as a watercolour. Its use is exemplified by the British artist JMW Turner. It was also mixed with Prussian blue or indigo to make the 18th-century watercolour, Hooker's green, a deep green colour for replicating foliage.

In more recent times, the supply of the resin has a darker side. Cambodian resin farmers collect gamboge in former battlefields. Unspent bullets are found mixed into the resin and unmarked landmines sometimes kill the farmers.

Unfortunately, like many organic colourants, gamboge fades rapidly in bright light and it now has very limited use in painting. Modern recipes replace gamboge with a lightfast synthetic yellow.

MYSTERIOUS COLOURS

MUMMY BROWN

THIS PIGMENT IS EXACTLY WHAT
ITS GRUESOME NAME IMPLIES.

Also known as mummia or *caput mortuum* (dead head), this dark brown pigment is made from the flesh, bones and wrappings of mummified Ancient Egyptian humans and animals.

It was originally used in Europe during the Middle Ages in the belief that it had medicinal qualities. This interpretation was based on the Ancient Greeks' medical use of bitumen. *Mumiya* was the Persian word for this black pitch, and it came to be applied to the bitumen that supposedly embalmed the shrouded corpses. Eventually, the word became attached to the preserved bodies themselves and the use of mummy brown by physicians for an incredibly wide range of ailments extended through to the 18th century.

In 1586, an English traveller called John Sanderson visited an ancient mass grave in Egypt and strolled among the corpses. He described how he 'broke off all parts of the bodies ... and brought home divers heads, hands, arms and feete for a shewe.' [6]

Mummy brown was first used in painting in the 16th century but it was most popular from the 18th to 19th centuries. A transparent rich brown colour, it was used in oil paint for glazing and shading. Greater understanding of the colour's grim source and increasing respect for the cultural importance of Egyptian artefacts led to general disapproval of its collection, sale and use in artists' colours. It fell out of favour and was virtually abandoned by the end of the 19th century.

VIII.

THE EXPLO[OF COLO[

OSION

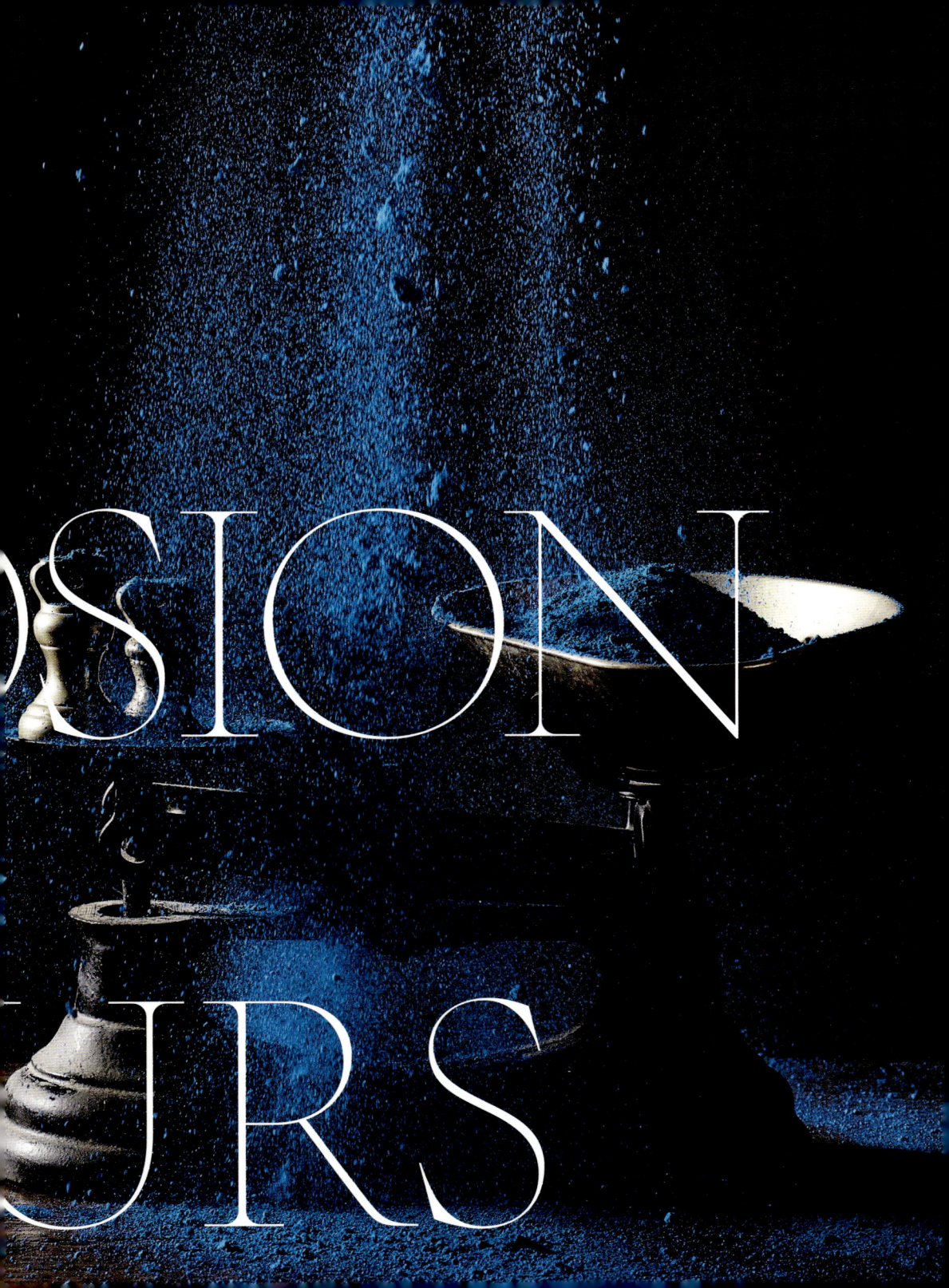

URS

THE EXPLOSION OF COLOURS

PRUSSIAN BLUE

CHANCE OFTEN PLAYS A ROLE
IN THE HISTORY OF ARTISTS' COLOURS.

Around 1704, the colour-maker Johann Jacob Diesbacht accidentally invented the first modern, artificially manufactured colour.

While working in his Berlin laboratory on a cochineal-based pigment called Florentine lake, Diesbacht ran out of alkali. He asked the alchemist Dippel, who shared the workspace, for some of the potash he had thrown away. What Diesbacht didn't know is that the potash had been contaminated with animal blood. After proceeding with his usual method, he was disappointed to find that his red lake turned out extremely pale. He attempted to concentrate it, only for it to turn first purple and then deep blue. The blood had triggered an unlikely chemical reaction that created the compound iron ferrocyanide, now known as Prussian blue.

Prussian blue was available to artists by 1724 and was immediately popular. It cost a fraction of the price of genuine ultramarine, making it a very attractive alternative. Its introduction also led to the demise of another mineral blue, azurite, from the artists' palette.

Prussian blue is easy and cheap to produce, non-toxic and intensely coloured. It is dark blue tending towards blue-black and, as a concentrated colour, has a bronze sheen. It has a high tinting strength and is lightfast, although it is sensitive to alkalis, which turn it brown.

Outside its artistic application, it has been used as a colourant to make blueprint paper, as a laundry blue, and in plastics, paper and cosmetics. There is even a pharmaceutical grade that is ingested to counteract radiation poisoning.

THE EXPLOSION OF COLOURS

LEAD CHROMATES

THESE PIGMENTS GIVE INTENSE
BUT FLEETING COLOURS.

The flame that shines twice as bright burns half as long.

This Taoist proverb applies poetically to the lead chromate pigments, not only because of their brief use by artists but also because of the often short life of their colour.

A deep orange powder is produced by crushing crocoite, a rare red mineral discovered in a Russian gold mine in 1770. Nine years later, the French chemist Louis-Nicolas Vauquelin deduced that this mineral was a combination of lead and a new element, initially called chrome (from *khrôma*, Ancient Greek for 'colour') and now known as chromium.

The possibilities for its use in pigment manufacturing were quickly recognised. By the beginning of the 19th century, the first methods of synthetic preparation from lead and chromium were invented. It could be made into strong bright colours that ranged from golden yellow through to rich orange and even red, offering intense colours that had never been available before.

Lead chromate pigments are relatively easy to manufacture and have excellent covering power but limited lightfastness and chemical stability.

Lead chromates were available by 1816 and employed liberally by artists, including Van Gogh. The effects of their discolouration are now highly evident in his work, as once-warm yellows have turned towards green. They were only used by artists for eighty years and were quickly replaced by the cadmium colours. By the end of the 19th century, they had virtually disappeared from artists' palettes.

THE EXPLOSION OF COLOURS

EMERALD GREEN

THIS DEADLY GREEN PIGMENT
CONTAINS COPPER AND ARSENIC.

Discovered in 1775 and named after its inventor the Swedish chemist, Carl Scheele, Scheele's green is an opaque yellow-green that was created to replace the historic copper-based greens of verdigris and malachite. Due to the paucity of greens available at that time, it was initially popular as an artists' pigment but it lost favour due to its toxicity and the fact that it discoloured in the presence of acids and sulphurs.

Emerald green (copper aceto-arsenite) was developed in 1808 in an attempt to improve Scheele's green. A more durable pigment than Scheele's green, emerald green still has a tendency to turn to brown when it is in contact with sulphur-containing colours, such as cadmiums or ultramarine. However, it was more brilliant than any previous green and was an instant favourite with dyers and artists.

Made by reacting verdigris with arsenic compounds, emerald green was, like its predecessor, extremely poisonous. Unfortunately, it was used extensively for printed wallpaper, which exposed its deadly nature. When the pigment reacts with moisture, it produces poisonous arsenic vapours. In damp climates, these vapours killed children in their nurseries. As early as 1815, suspicion fell on the colour but it took decades for its use in household items, and even food colouring, to be prohibited.

Emerald green was also known as Schweinfurt green, Veronese green and Vienna green – in fact, more than eighty different names have been found for this pigment. The proliferation of terms was probably used to disguise its toxic infamy. Even with scientific proof of its highly toxic nature, the production of emerald green paint was not banned until the 1960s.

THE EXPLOSION OF COLOURS

COBALT

COBALT IS NAMED AFTER
A MALICIOUS GOBLIN.

In German folklore, *kobold* was a sprite that haunted and tormented the miners, and was believed to be responsible for the poisonous nature of the mines.

Cobalt occurs naturally in the ore smaltite, which is a mixture of cobalt and nickel arsenides. Found in silver mines, smaltite forms a brilliant blue crystal that contains deadly arsenic. It was known by the miners as 'cobalt bloom'.

Cobalt had been used as a colouring agent in pigments and ceramic glazes since antiquity, but the cobalt colours created in the 19th century were far more intense and stable. The French chemist Louis Jacques Thénard synthesised modern cobalt blue in 1802. It was purer in colour than virtually all the blues that preceded it, and was immediately adopted by artists.

Cobalt has chameleon-like qualities: in combination with other elements it can be made into green, violet and yellow. Cobalt green's composition is similar to cobalt blue, but the alumina that gives cobalt blue its distinctive colour is replaced with zinc oxide. Cobalt green was discovered in 1780, but it wasn't until 1835 that the industrial production of zinc oxide made its manufacture commercially practical.

Cobalt violet was manufactured from 1859 but its high cost and low tinting power limit its use among painters. Cobalt yellow, known to artists as aureolin, was first synthesised in 1831 but was not available as a pigment until the 1850s.

Although they only have modest tinting strength, the cobalt colours offered artists brighter hues than ever before. The Impressionists and Post-Impressionists used them enthusiastically, to dazzling effect.

THE EXPLOSION OF COLOURS

POTTER'S PINK

THIS PIGMENT WAS INVENTED BY AN ANONYMOUS STAFFORDSHIRE POTTER.

Also known as tin pink, *nelkenfarbe* (carnation colour) or pinkcolour, potter's pink was used at the beginning of the 19th century as a pigment for ceramics.

Another metal-oxide-based pigment, potter's pink is manufactured from tin oxide roasted with chalk and chromium oxide and was one of the first stable modern synthetic colours. As an artists' pigment it was mostly employed for watercolour as a lightfast alternative to the madder lakes, where the problem of colour fading was still an issue. A semitransparent, dusky, deep pink colour with limited colour strength, this is especially obvious when used in oil paint.

Pink is unusual as a colour because, although it is only red tinted with white, we recognise it as a colour in its own right. Blues and yellows with similar additions of white are simply described as pale versions of their original hue. Maybe its approximation with flesh gives this tone its individual power.

Like many of the earliest synthetic pigments of the 18th and early 19th centuries, it fell out of favour when brighter, more powerful and cheaper alternatives were introduced. It is, however, still available. Its subtlety, compared to 20th-century pigments, offers artists a softer tone and its application in landscape painting attracts great interest. Perhaps this historic pigment will make a comeback in the 21st century.

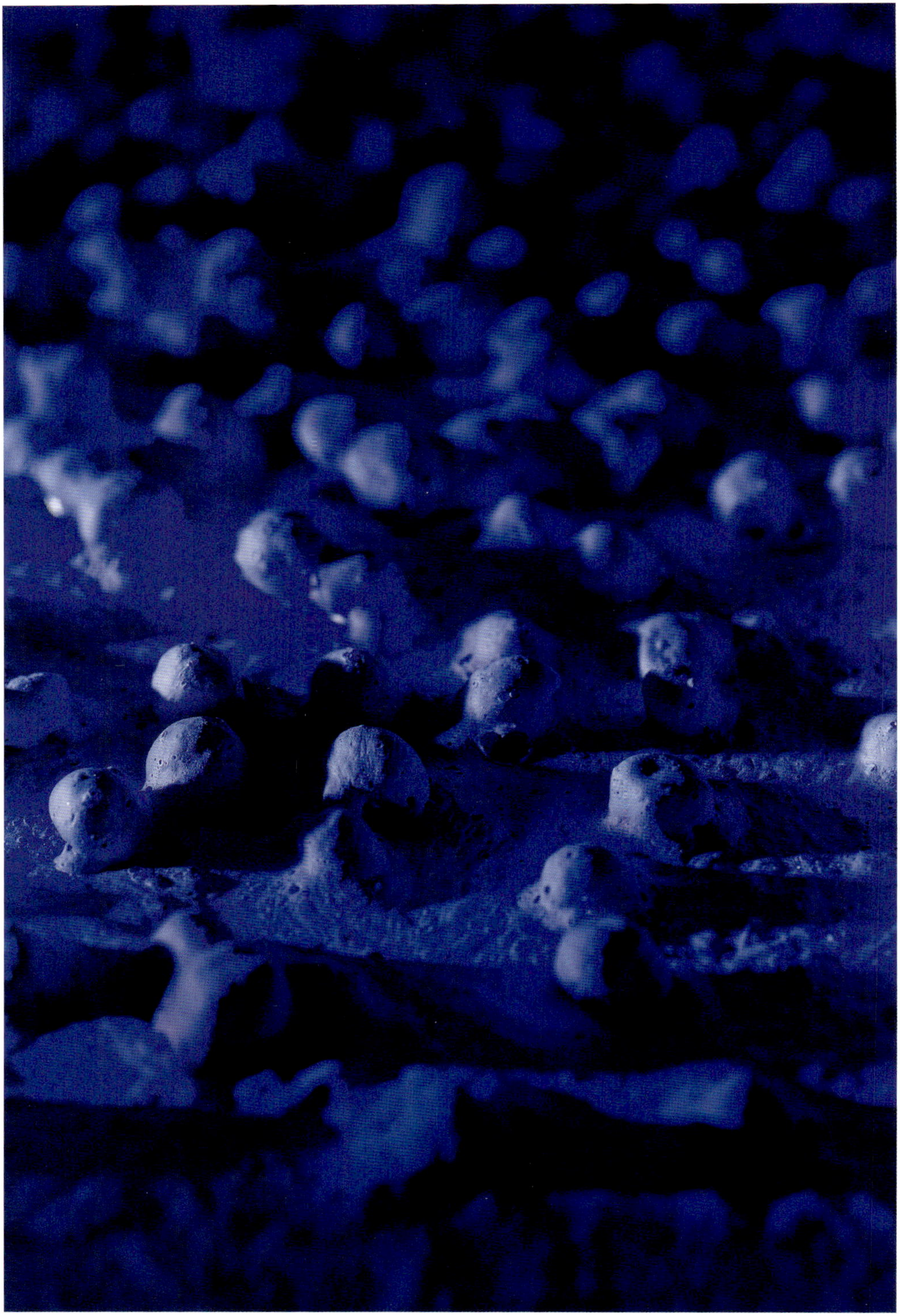

ULTRAMARINE

GENUINE ULTRAMARINE WAS SO EXPENSIVE
THAT ARTISTS COULD RARELY AFFORD IT.

In 1824, the Société d'Encouragement offered a prize of 6000 francs to anyone who could produce a synthetic replacement with a cost that did not exceed 300 francs per kilogram.

In the previous century, it had been noticed that glassy blue deposits were collecting on the walls of lime kilns. These deposits were used locally as a substitute for lapis lazuli in decorative work. From these observations, chemists began their investigations.

In 1828, Jean-Baptiste Guimet perfected a method of producing artificial ultramarine by heating a mixture of china clay, soda ash, coal, charcoal, silica and sulphur. Because the French government had requested its invention, it was known as French ultramarine to differentiate it from the natural product.

The terms ultramarine and French ultramarine are now used by paint-makers to describe different blue hues. The names are an echo of a time when the natural and artificial pigments were still sold together.

As well as the famous blue pigment, ultramarine can be carefully manufactured into redder shades. Ultramarine violet is made by heating a mixture of ultramarine blue and ammonium chloride. Ultramarine pink is derived from ultramarine violet by heating it with gaseous hydrochloric acid. However, as the ultramarines move towards the red shades, they lose their tinting strength and opacity.

The ultramarines are permanent, non-toxic and cheap to produce. Artists can now afford to employ them with an abandon that the Renaissance painters would have found impossible to imagine.

THE EXPLOSION OF COLOURS

CADMIUM

THE IMPACT OF CADMIUM COLOURS
ON ART CANNOT BE UNDERESTIMATED.

What would later 19th-century art have looked like without the invention of cadmium colours? The Impressionists devoured these previously unseen bright opaque primaries and used them to create vibrant, colour-charged interpretations of the world around them.

The name cadmium comes from the Latin *cadmia*, which is a traditional term for the zinc ore from which cadmium is extracted. Far more stable than the lead chromates they replaced, cadmium colours have excellent hiding power, are very lightfast and were extremely bright compared to previously available colours.

Cadmium yellow is made from cadmium sulphide. Although it was first discovered in 1817, commercial production of cadmium yellow pigments was delayed until 1840 because of the scarcity of the metal ore. They can be made into various shades, ranging from a lemon yellow through to a bold primary yellow and even deep oranges, depending on the size of the grains produced in the manufacturing process.

Cadmium red is cadmium sulphide with the addition of selenium, an element poetically named after *selḗ nē*, the Ancient Greek word for moon. Cadmium reds were introduced in 1910 and range from warm light reds through to red-purples. They quickly replaced the toxic pigment vermilion.

Cadmium pigments have enormous applications in industrial processes and their use in colouring plastics is particularly significant. They are still incredibly important for artists to this day because of their versatility and stability.

THE EXPLOSION OF COLOURS

CERULEAN BLUE

CERULEAN BLUE APPROXIMATES
BRIGHT, CLEAR SKIES.

A mixture of cobalt and tin oxides (cobalt stannate), cerulean blue derives its name comes from the Latin *caeruleus* (dark blue *caelum*). *Caelum* translates as 'the vault of heaven'. In France, it is called *bleu céleste* (heavenly blue).

In classical times, *caeruleum* was a general term used to describe many blue pigments. The Roman author Pliny states there was a diverse range of caeruleum, although his descriptions are actually for variations of Egyptian blue, which was still in production at the time of his writing. The name cerulean was, for a time, also applied to mixtures of copper and cobalt oxides, such as natural azurite and synthetic smalt.

The Swiss chemist Albrecht Höpfner developed a prototype of cerulean blue in the late 18th century but, curiously, it was forgotten. It was rediscovered in the 1860s with a slight improvement in the manufacturing process. Cerulean blue has a slight green hue, which makes it the perfect foil for the red tone of cobalt blue. It became a useful addition to the other modern blue pigments that were becoming popular among artists in the mid-19th century. Upon its release, the Impressionists immediately employed it extensively. It is very stable and does not react to light or chemicals, making it permanent and invaluable for artists.

There is a second version of cerulean blue in which the tin is replaced by chromium, which produces a greener colour. This is often known as cobalt turquoise. Both pigments are still of great importance to artists today.

THE EXPLOSION OF COLOURS

MANGANESE

'I HAVE FINALLY DISCOVERED THE TRUE COLOUR
OF THE ATMOSPHERE, IT'S VIOLET …'

This is what Claude Monet proclaimed in 1881. The Impressionists so adored the new hue that critics accused them of having 'violettomania'.

Discovered in 1868, manganese violet has played an important role in modern art. Relatively cheap to produce, it is an opaque red-violet colour that quickly replaced the weaker cobalt violet. Although it has only modest pigment strength, it gives exquisite light-filled violets. This perfectly matched the Impressionists' theories that shadows were not black but coloured by the complementary hue of the light striking the subject.

Manganese blue is a clear azure blue. First discovered in 1907, it only went into commercial production in the mid-1930s. Although it was initially sold for colouring cement, its potential as an artists' pigment was quickly recognised. However, as it is a relatively weak tint, it never became an important pigment. Its production was discontinued in the 1990s due to concerns about its environmental toxicity.

The element manganese lends itself to a wide variety of colours. By definition, manganese oxide must be present in umber and sienna pigments – their colour comes directly from its inclusion in their composition.

Manganese black, also known as mineral black, can be found in nature as the mineral pyrolusite although it is generally used in its synthesised form. It has a brown undertone and, like all manganese-based pigments, acts as a drying agent for oil paints.

IX.
A BRA[VE]
NEW W[ORLD]
WORL[D]
COLO[UR]

VE
D OF
JR

A BRAVE NEW WORLD OF COLOUR

MARS COLOURS

ALL YELLOW, ORANGE AND RED
NATURAL EARTHS ARE PREDOMINANTLY
COMPOSED OF IRON OXIDES.

The method of manufacturing synthetic iron oxide was known from at least the 15th century, although large-scale production did not begin until the middle of the 19th century.

Historical manufacturing methods included the simple heating of iron filings by dissolving them in *aqua regia* (a mixture of hydrochloric and nitric acids) and roasting the resulting iron salt, or by heating *martial vitriol* (iron sulphate) directly.

From the mid-18th century, these synthetic pigments were called Mars colours. The name is a literal translation of *crocus martius*, the Latin name alchemists gave to artificial pigments made with iron oxide.

With the rise of industrial chemistry in the 18th century, materials that were previously unattainable became both affordable and readily available. The use of sulphuric acid as a bleach for textiles made commercial production of Mars pigments possible.

Modern manufacturing methods allow the colour to be precisely engineered to the desired hue. The purest and finest oxides are produced from the precipitation and hydrolysis of iron salt solutions. Hue and tinting strength are affected by hydration, particle size and the presence of additives such as manganese.

Because the manufacturing processes for iron oxide pigments can be controlled so precisely, the modern versions are purer, have smaller particle sizes, greater tinting strength and are much more opaque than natural ochres.

A BRAVE NEW WORLD OF COLOUR

ZINC WHITE

THIS NON-TOXIC ALTERNATIVE TO
LEAD WHITE WAS SORELY NEEDED.

By the end of the 18th century, the risks associated with lead white were a matter of serious concern. Lead white was made in huge amounts. It was the only white pigment in wide use, appearing not only in artists' materials but also as a colourant in many industrial applications.

The modern method of zinc white production, known as the French process, was perfected in 1844 and the pigment was rapidly adopted. It was superior to lead white because of its lack of toxicity. It was also cheaper to produce.

Zinc was well known to the Ancient Greeks, who observed fluffy deposits of condensed zinc oxide. Zinc metal was first isolated around the 10th century in India and China. The process was later adopted in Persia, where zinc was called *tutiya* (smoke), because it looks like white vapour when the zinc ore is smelted. During the 18th century, zinc did not have a universally accepted name. The metal was called *tutanego* and zinc oxide was known as *tutty*.

Although known since the Middle Ages as 'flowers of zinc', it was rarely used as a pigment until 1834 when it was introduced as a watercolour pigment called Chinese white.

In artists' paints, zinc white has a cold, flat tone compared with lead white, and it doesn't have the same opacity as titanium white. It is, therefore, often recommended as a mixing white as it does not overpower the colours with which it is blended.

A BRAVE NEW WORLD OF COLOUR

TITANIUM WHITE

THE BRIGHTEST OF WHITES IRONICALLY
COMES FROM A BLACK MINERAL.

The dark-coloured mineral ilmenite is rich in iron and titanium and when titanium was identified in 1791, its potential uses as a pigment began to be explored. The earliest uses of synthetic titanium dioxide in the late 19th century were to add opacity and increase acid resistance to ceramic glazes.

Naturally occurring titanium was not used as a pigment because it was too highly contaminated with iron. A suitable purification method was not developed until the 1920s. First treated with sulphuric acid to form a solution of sulphate, it is hydrolysed to form a white precipitate of hydrated titanium dioxide. This is roasted in a furnace to form the white pigment.

Acceptance of this new pigment was slow. The paint industry was reluctant to introduce it as it was in limited supply and the cost was higher compared to the commercially dominant lead white. However, the introduction of laws in the 1920s that restricted the use of toxic lead-based pigments made titanium white an obvious choice for the modern age.

Nowadays, titanium dioxide is manufactured using a chloride process. It is very safe to use, has the highest hiding power of all the white pigments, has excellent lightfastness and is very popular in contemporary artists' colours. It has a wide range of applications, including decorative paints, plastics and printing inks. It is the most widely used pigment of all time.

SYNTHETIC CHEMISTRY OF THE MODERN WORLD

A BRAVE NEW WORLD OF COLOUR

Large commercial industries have always driven colour innovation. For example, the investigation into the chemistry of textile dyes during the industrial revolution led to a rapid expansion in colour for paints.

After William Perkin accidentally synthesised mauve in 1856, many chemists explored the chemical construction of colours in the hope of making their fortune too. They looked to coal tar, the thick black residue left over from the extraction of gas from coal. Coal tar is composed of hydrocarbons (organic compounds of hydrogen and carbon), which combine with other elements to form new synthetic materials.

The very first synthetic organic pigment was tartrazine yellow, which was patented in 1884. It was made from azo yellow dye and is still in use as an artists' pigment. Soon after, a vast numbers of new azo dyes and pigments were introduced, including the arylide group. They ranged from a clean, bright yellow to lemons and warm oranges. The arylides were commercially available from 1925, but only became widely used after World War II as a replacement for cadmium yellow.

Most synthetic organic pigments are part of the polycyclic colour group that includes phthalocyanines, quinacridones, naphthols, perylenes, anthraquinones, dioxazines and pyrroles. They are now abundantly used in painting, although the artist may not be aware of this. Their chemical names are often replaced with trade names such as Monastral or Hansa, or performance-describing names like 'permanent'. They are often described by their colour equivalence to more traditional pigments. For example, cadmium yellow 'hue' is actually made with azo yellow pigment.

The phthalocyanines, aka phthalo (pronounced 'thalo') pigments, were introduced in the early 20th century. They were the first organic colourants to be called 'true' pigments. Until then, all pigments based on organic material – such as madder and cochineal – had started as a dye or lake before being converted into an insoluble pigment.

The phthalo pigments have enormous tinting strength, lightfastness and chemical resistance. They quickly dominated the blue and green pigment markets in all applications. They are immensely important for artists due to their intensity and colour purity. Phthalo blue can be almost pure chromatic blue, with no bias towards green or red, while phthalo green is a deep, cool, emerald green.

SYNTHETIC CHEMISTRY OF THE MODERN WORLD

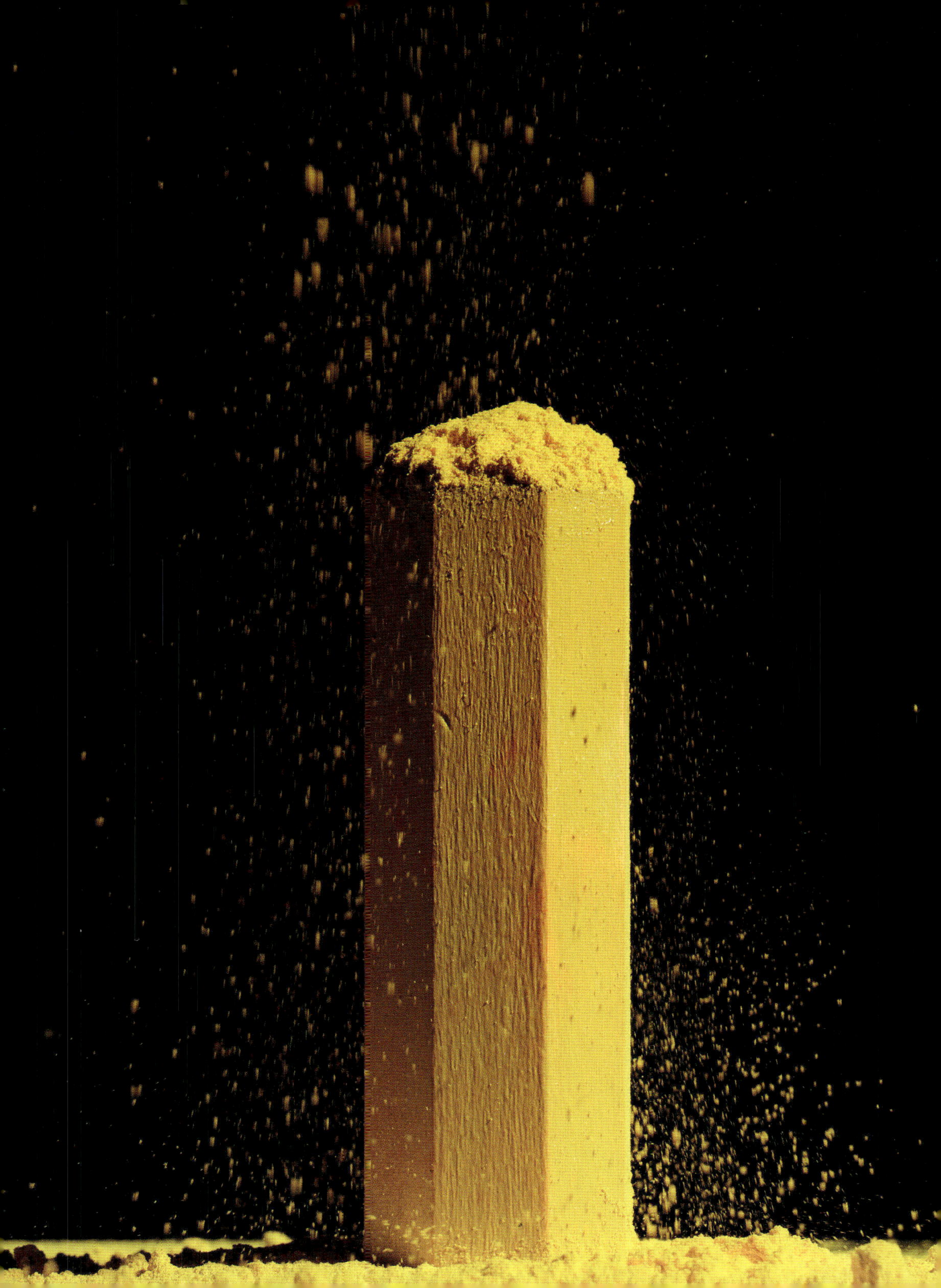

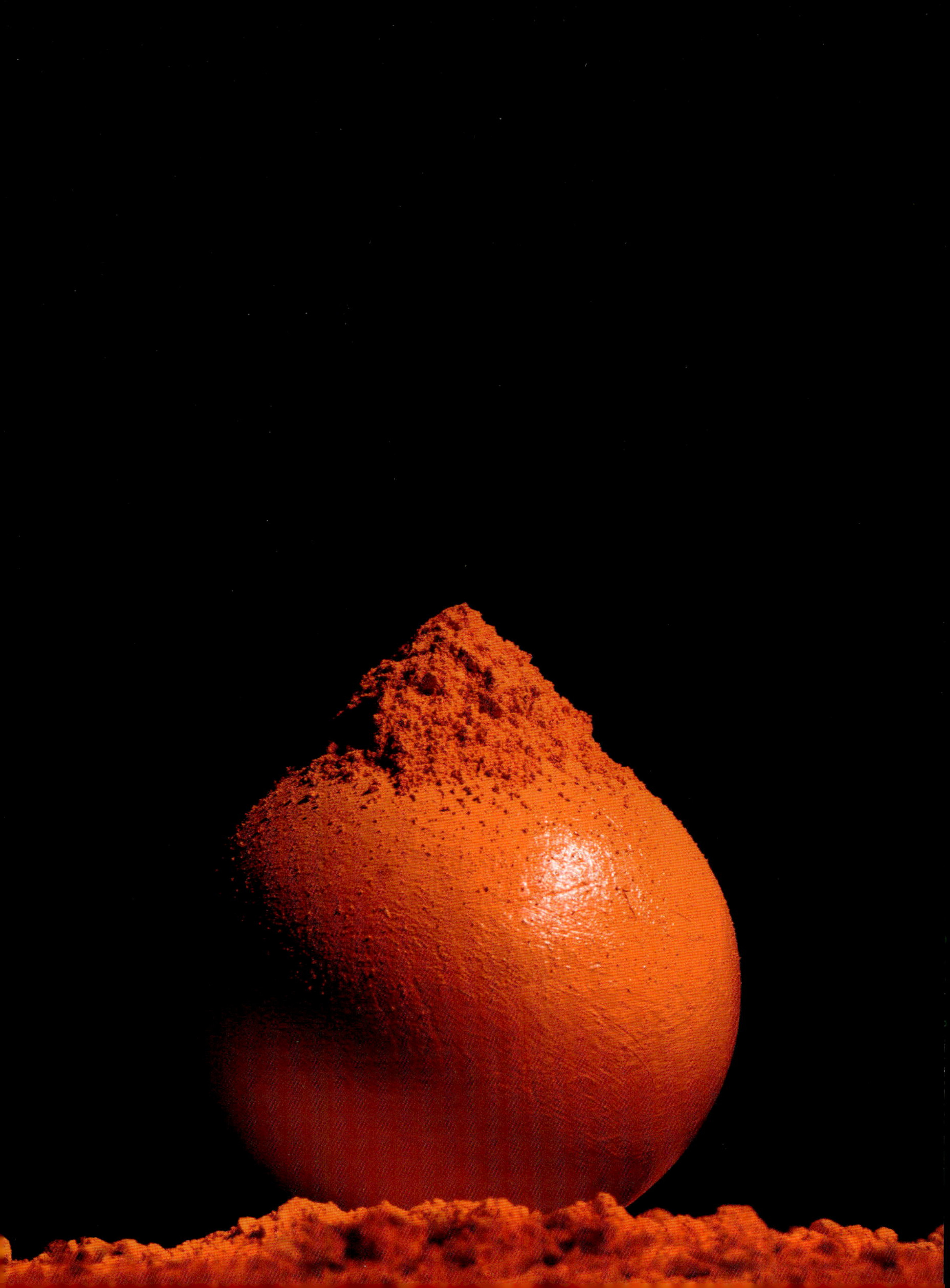

Another important group of synthetic pigments are the quinacridones, which range from yellow-reds through to violets. Evidence of the quinacridone structure was first uncovered in 1896 but the pigment was not synthesised until 1935. It took more than twenty years for commercial progress to be made, and it was not until 1953 that the synthesis was perfected. Used in artists' paints for the last fifty years, they offer exceptionally clean, translucent colours that are perfect for glazing and creating pure colour mixes.

Naturally transparent pigments have limited use as body colour; however, recent developments have led to new highly opaque pigments. Pyrrole red, also known as Ferrari red, was first seen on all red Ferraris in 2000. It is based on the organic compound diketo-pyrrolo-pyrrole (DPP) and was first synthesised in 1974. Both the red and its orange-shade cousin share important qualities for artistic use: they are opaque, extremely lightfast, chromatically very pure, and non-toxic.

The extensive range of organic synthetic pigments is always growing, satisfying demands for high-performance colour and pushing the boundaries of performance.

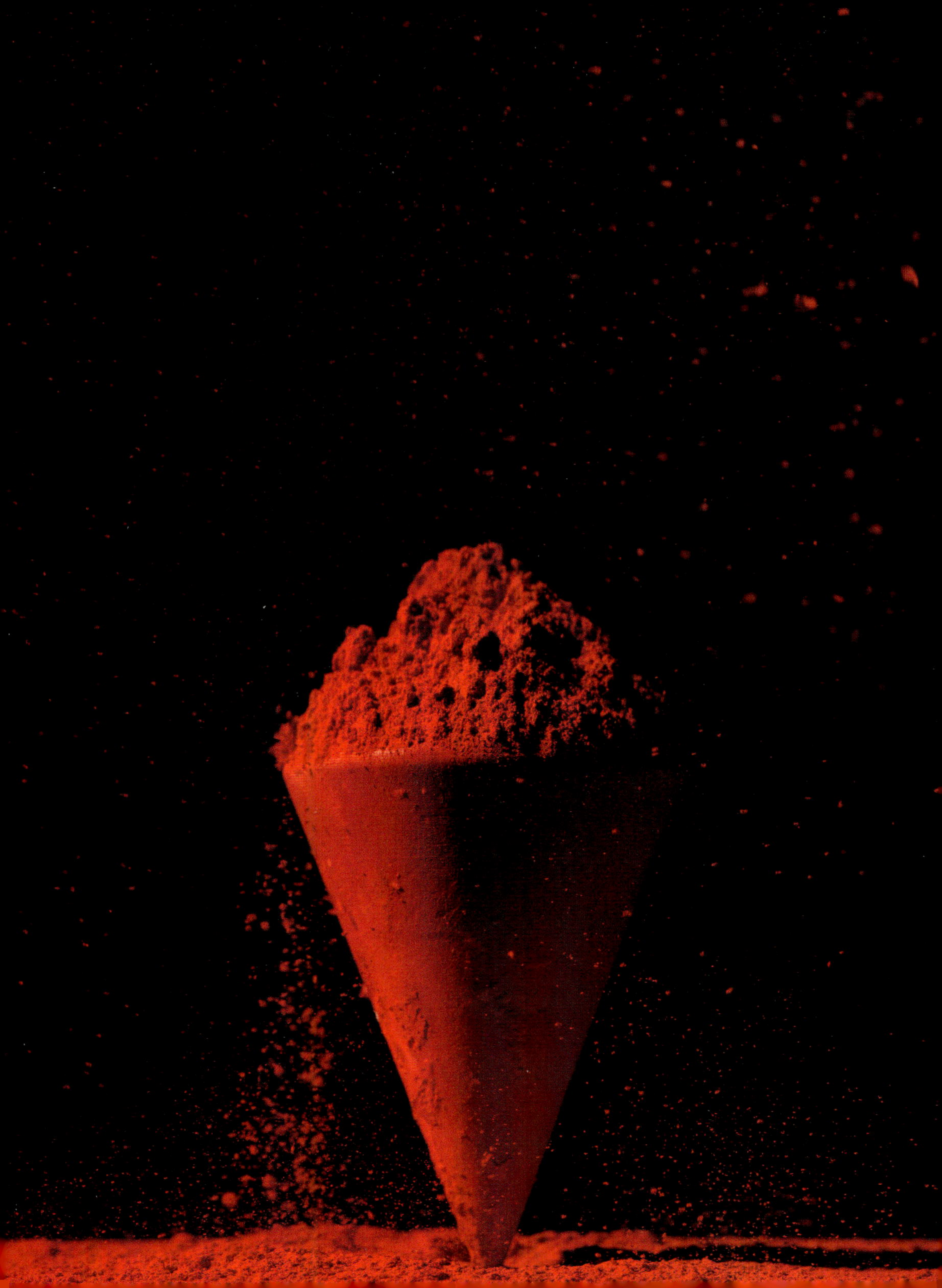

X.

THE
SCIEN
MODE
COLO

CE OF
RN
UR

THE SCIENCE OF MODERN COLOUR

FLUORESCENCE

FLUORESCENCE IS ALL ABOUT ENERGY!

When light shines on a fluorescent pigment, the molecule absorbs energy as a photon of light. This excites the electrons inside the molecule. As the excited electrons lose energy, they emit the absorbed photon of light. We see this as fluorescence.

Fluorescent colours use a larger amount of both the visible spectrum and the lower wavelengths compared to conventional colours. As a result, they are perceived as far more intense colours. Conventional pigments can reflect a maximum of ninety per cent of their colour; a fluorescent pigment can reflect three times as much.

In nature, fluorescence can be observed in rubies and emeralds. Many animals exhibit biofluorescence, including sharks, mantis shrimp, amphibians and parrots.

Most fluorescence only occurs under ultraviolet light, and the colours stop glowing as soon as these invisible wavelengths are no longer present. However, in the 1940s the American Switzer brothers invented daylight fluorescent pigments. They called them Day-Glo. These pigments are luminescent, which means they glow purely by being struck with visible light from the sun. They were extensively used during World War II for high visibility markings and signalling, but later made their way into safety equipment, advertising and packaging.

Fluorescent pigments also made their way into art. The psychedelic paintings of the 1960s took advantage of their other-worldly incandescence and although they are completely fugitive and not lightfast, their unique qualities are still attractive to artists today.

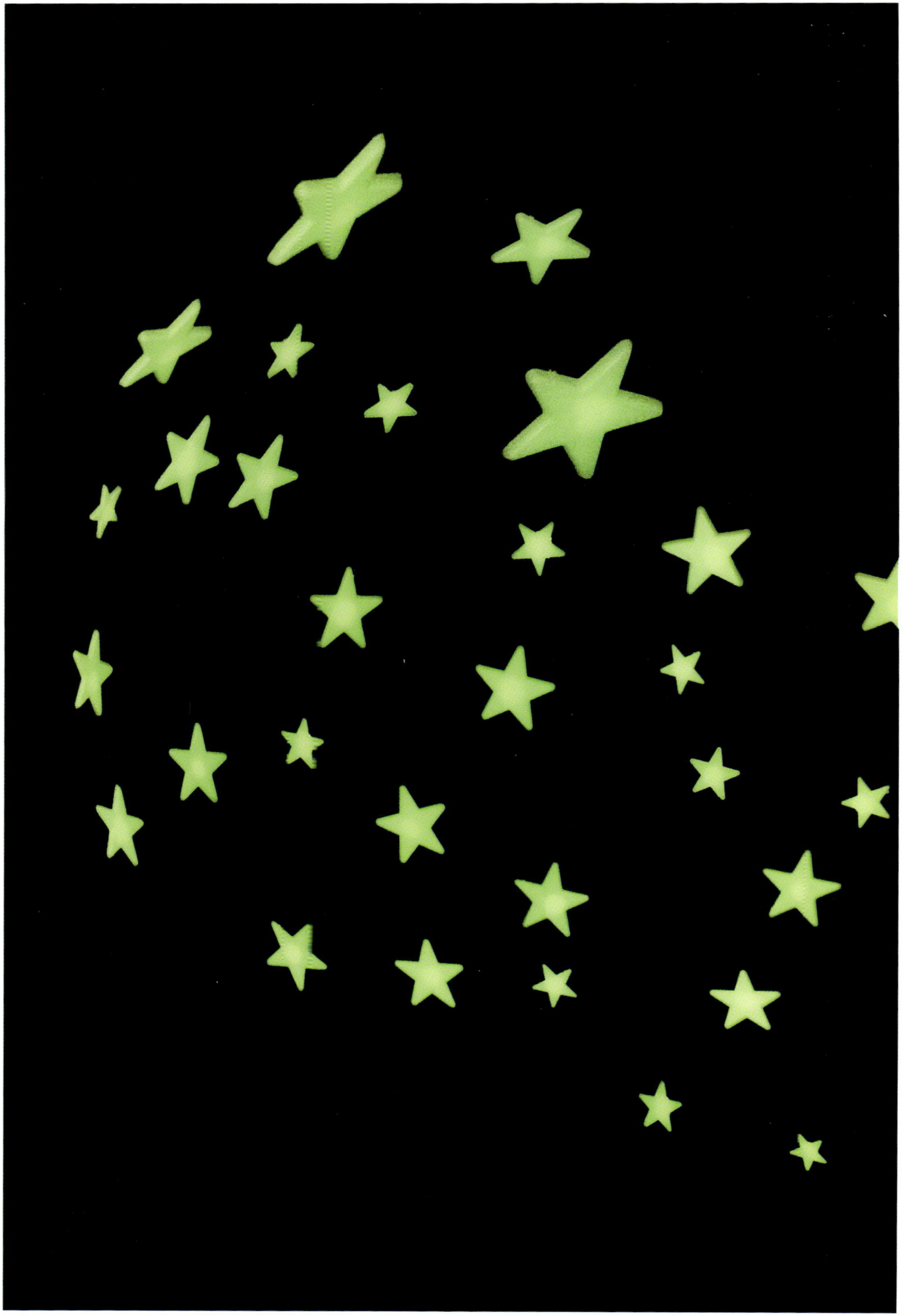

PHOTOLUMINESCENCE

PHOSPHORUS MEANS 'THE LIGHT BEARER'.

Have you ever seen stars on the ceiling of a child's bedroom that appear to be glowing in the dark? This is a phosphorescent glow-in-the-dark pigment that works by absorbing light as energy and using it to excite phosphor molecules. Unlike fluorescent pigments, which stop glowing as soon as the light source disappears, phosphorescent pigments continue to emit light for some time after.

Phosphorus was the name the Ancient Greeks gave the planet Venus. The appearance of Venus in the morning sky heralded the imminent sunrise. Since the Middle Ages, the term phosphor has been used for materials that glow in the dark after being exposed to light.

The first commercially developed glow-in-the-dark pigment was a radioluminescent pigment invented in 1908. It was used in phosphorescent paint for watch faces and compasses to allow them be seen in the dark. It was also extremely dangerous, as it releases radioactive radium. The paint was eventually withdrawn from use after it poisoned the factory workers who were working with it.

In the 1930s, zinc sulphide replaced radium as the base chemical for phosphors. It was safer, but it only emitted a glow for a short period of time. Zinc sulphide was superseded by strontium aluminate, which can produce many hours of continuous glow.

Many important applications have been developed based on photoluminescence. They include safety signs, bank notes, toys and paints. Though not completely fugitive like fluorescent pigments, its ability to be recharged and re-emit light weakens gradually over a period of decades.

THE SCIENCE OF MODERN COLOUR

YInMn BLUE

YInMn BLUE IS NAMED FOR ITS CHEMICAL MAKEUP: YTTRIUM, INDIUM AND MANGANESE.

Like Prussian blue, this blue pigment was discovered by accident. In 2009, the chemist Mas Subramanian and his students at Oregon State University were investigating new materials that could be used for manufacturing electronics. A graduate student noticed that one of their samples turned a bright blue colour when it was heated, to which Subramanian responded, 'Luck favours the alert mind'. This was an approximation of an 1854 quote from the French chemist Louis Pasteur: 'In the fields of observation chance favours only the prepared mind'. In 2016, an announcement of its invention in preparation for production was released.

YInMn blue is durable, safe and easy to produce, although the elements needed to create it are rare and expensive. It has been suggested one major commercial application for YInMn blue could be as a roof colourant to reduce heat penetration, due to its low thermal conductivity. But, just like the introduction of new pigments in the 19th and 20th centuries, colours that are developed for industry may well become a part of the artists' palette in the future.

The original chemical structure of YInMn Blue seems to allow for its manipulation into a wider range of coloured pigment. Replacing indium with zinc and titanium creates a violet pigment. To make yellows and oranges, iron is added to the composition, and greens are produced with the addition of copper.

THE SCIENCE OF MODERN COLOUR

VANTABLACK

THIS IS ONE OF THE STRANGEST
PIGMENTS EVER CREATED.

Vantablack is made of a forest of vertical tubes that are grown on a surface. Incredibly, it is the darkest material on the planet. Vantablack is an acronym of Vertically Aligned NanoTube Arrays. Made by a process of chemical vapour deposition, it absorbs up to 99.96 per cent of all visible light.

When light hits vantablack's surface, instead of bouncing off as would happen with a standard colour pigment, it becomes trapped and deflected among the structure of evenly spaced nanotubes, eventually being almost completely absorbed. The fraction of light the coating does reflect comes from light particles that hit the very tops of the tubes.

On a flat surface, it initially looks like an incredibly dark black, but when applied to a three-dimensional object, the form almost vanishes into a two-dimensional silhouette.

Nanotubes are incredibly small. One square centimetre contains approximately one billion nanotubes, even though more than 99 per cent of the material is empty space. Its delicacy means that it cannot be touched without damaging the coating. It would collapse under the weight of contact.

Vantablack's development has been driven by scientific applications in which the surface won't be disturbed by physical contact. When used in telescopes in space, its function is to absorb any stray, unwanted light that might interfere with the instruments.

Because of its fragile nature, it seems unlikely that it can be used in traditional artists' paint. But, as the history of colour has shown us, innovation has always forged a new, more colourful path.

MAKING COLOUR VISCERAL: HOW PAINT IS MADE

THE SCIENCE OF MODERN COLOUR

This book has looked at the origins of historical and contemporary pigments, but pigments are hardly ever used in their raw form. To be usefully employed as a colour, billions of individual grains of pigment must be glued together with a binder. This is, in essence, how you make paint.

Throughout history, people have found ways to permanently 'fix' colour to create lasting images of the most exquisite beauty. For instance, the binding of pigments in Neolithic cave paintings was probably serendipitous; cave walls containing silicas or limestone trapped the pigment and locked it to the surface over time. Since then, we have discovered a host of sticky, adhesive materials in nature that could hold pigments in place. Some of these earliest binders are still used by artists. Gum arabic, the water-soluble sap of the North African acacia tree, makes watercolours; and beeswax, collected and refined from hives, makes encaustic (molten wax) paint.

Mixing pigments with different binders successfully converts them into a material for uses as diverse as house paints, plastics, writing inks, automotive coatings, paper and – of most interest to me – artists' paint.

In my role as a master paint-maker, I make oil paint, which are by dispersing pigments in a 'drying oil' such as linseed, walnut, poppy or safflower oils. Linseed oil is by far the most important and widely used drying oil. When drying oils absorb oxygen they convert from a liquid into a hard, permanent coating. Pigments can be bound with very small amounts of oil. This means that oil paints contain much higher amounts of the pigment than watercolour or acrylic paints. For artists, this gives the paint a physical feeling. The paintbrush is literally pushing around dense, coloured pastes.

So how do we make our paint? Our first task was to source a high-quality linseed oil. We selected ours after sampling dozens of products from suppliers all over the world. We were looking for a clean, straw-coloured oil that was free of natural impurities. It had to have a good drying rate and minimal yellowing as it aged. Eventually we chose exceptional bright, clear oils made in Holland and Germany.

Next comes the selection of the pigments. There are so many manufacturers of pigments that the choice seems overwhelming. We hunt out pigments that have qualities equal to their noble intended use: they must be as lightfast as possible, chemically stable and exhibit colour qualities of benefit to the artist. The vast majority of pigments do not meet our needs. They are built for larger, more commercially important industries and have been tailored for industrial applications.

To select our pigments, we go through a long period of investigation. We select colours of interest, research the chemical construction of the pigment, and assess its suitability for artists' paint before requesting samples for laboratory trials. The anticipation of opening a sample box and seeing a new pigment for the first time, in its raw unadulterated form, is exhilarating. There is always the nervous hope that the promise held out by this new pigment will be borne out, that its potency will not dull, and that its colour will not be lost when it is mixed with the binder. Backwards and forwards go the experiments – working out the right amount of pigment to add to the oil and correcting for undesirable qualities. Like a chef honing a new dish, small, delicate changes in the recipe can lead to dramatic differences in the finished product.

When we are ready to make the paint, the linseed oil is weighed out into 60-litre heavy-duty stainless-steel bowls. All of our manufacturing equipment and surfaces are stainless steel. The equipment is kept meticulously clean to prevent any chance of other colours contaminating the purity of each new batch.

Stearate, a wax-like material that is essential to the wetting and stability of the paint, is weighed and added to the oil. The bowl is secured in a planetary mixer and large, powerful motors slowly rotate the blade through the wax and oil mixture.

Next, another steel bowl is placed on the electronic scales, ready for the pigment. Even after all these years, opening the bins of pure pigment is a ridiculously breathtaking assault on the eyes. The pigment is scooped out, weighed and added slowly to the oil. There are no short cuts. Adding all the pigment at once would make incorporation impossible. The liquid oil allows the individual grains of colour to slide over each other. The physical shape of pigments means that, without this lubrication, they would drag over each other, causing extraordinary resistance, reducing the mixing action and – as happened once very early on – breaking the very expensive blade of the mixer.

The slow churning of the paste begins. Over the rumbling of the mixer's motor, you can hear delicious slurping noises as the blade methodically drives through the mixture. As the dry pigment is gradually incorporated with the wet oil, it changes from an incredibly thick batter into what looks like an enormous vat of vividly coloured butter.

This process can take as long as four hours, but it is not the finished paint. Under close inspection, vast quantities of the pigments still cling together rather than being individually coated. This is where the triple-roll mill comes in.

A triple-roll mill is at the heart of paint-making. At its most basic, it is three horizontal granite rollers that each run at different speeds and spin in alternating directions. The paste is scraped out of the mixer's bowl with a baker's blade and dropped into the hopper. Each giant dollop makes a delicious slap as it plops onto the rollers below. The paste is drawn down into the tiny space between the rollers, again and again. With each pass, the space is narrowed to more aggressively separate the pigment particles. If you have ever used a pasta-making machine with its two rollers forcing the dough through the small space between them you can understand the paint-making process. Just as the roughly made dough cannot pass through the narrowest setting first, so the pigment-paste must be passed through the mill rollers multiple times. It's just that our mill is like a pasta-machine on steroids, with three rollers rather than two and a massive motor to drive the material through. For soft pigments such as zinc white only three passes are needed, but the synthetics can take up to nine passes. Synthetic pigments are very difficult to prize apart: their incredibly small size and specific shape mean they have to be painstakingly teased into dispersion.

The paint-maker must be constantly attentive to the vagaries of milling. Rollers heat up under the friction of pigment particles, which alters the size of the roller gap, and the fluidity of the oil is affected by changes in ambient temperature. Also, pigments behave differently from one batch to another. This is especially true of the natural earths, which vary in their mineral make-up depending on the part of the seam the earth was dug from.

Towards the end of the paint-making process, we take samples of the paint and test it for quality. Historically, paint-makers would rub the paint between their thumbnails – a simple but surprisingly delicate solution to feel for the grittiness of unmixed pigment. Nowadays we use a precisely honed stainless-steel gauge to check the quality of dispersion.

But we are still not ready to sign off on the product. Two extremely thin films of the freshly made paint are applied to paint-maker's cards. One daub is the pure paint. The other is the paint mixed with a specified amount of titanium white. By placing the card next to one from a previous batch of the same colour, we can ensure that every time we make the paint it has identical colour, tinting strength, tint colour and undertone to all previous versions.

Only after the paint has passed these tests is it approved for packing. It is hand-filled into collapsible aluminium paint tubes, labelled with hand-painted swatches of the individual colour, boxed and shipped to studios around the world.

TURN PAST 6 THEN SET FOR
0 1 2 3 4 5 6 8 10 12 14 16 20 22 24 26 28 30

STOP MIXER
BEFORE
CHANGING SPEED
1 2 3 4 3

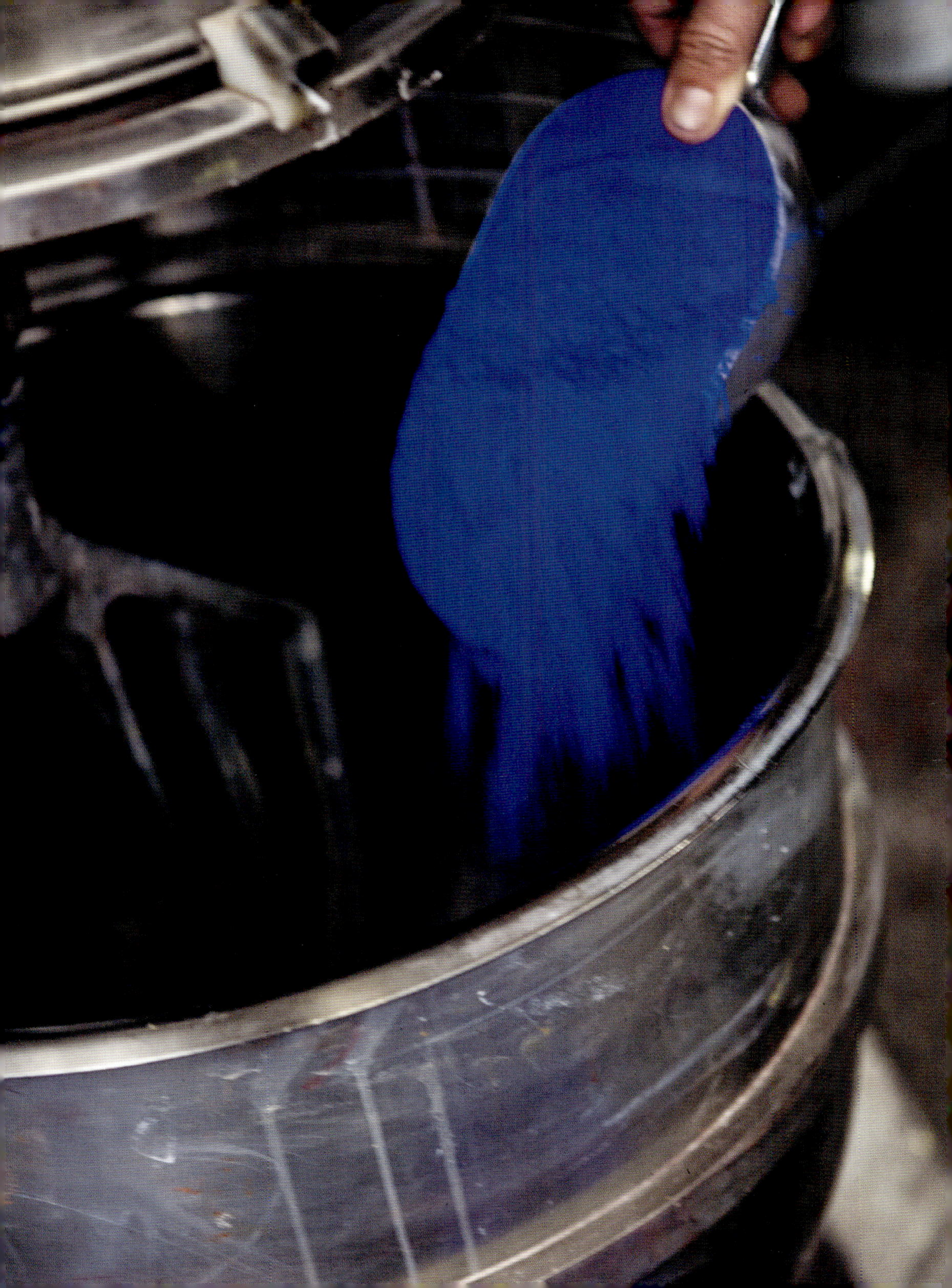

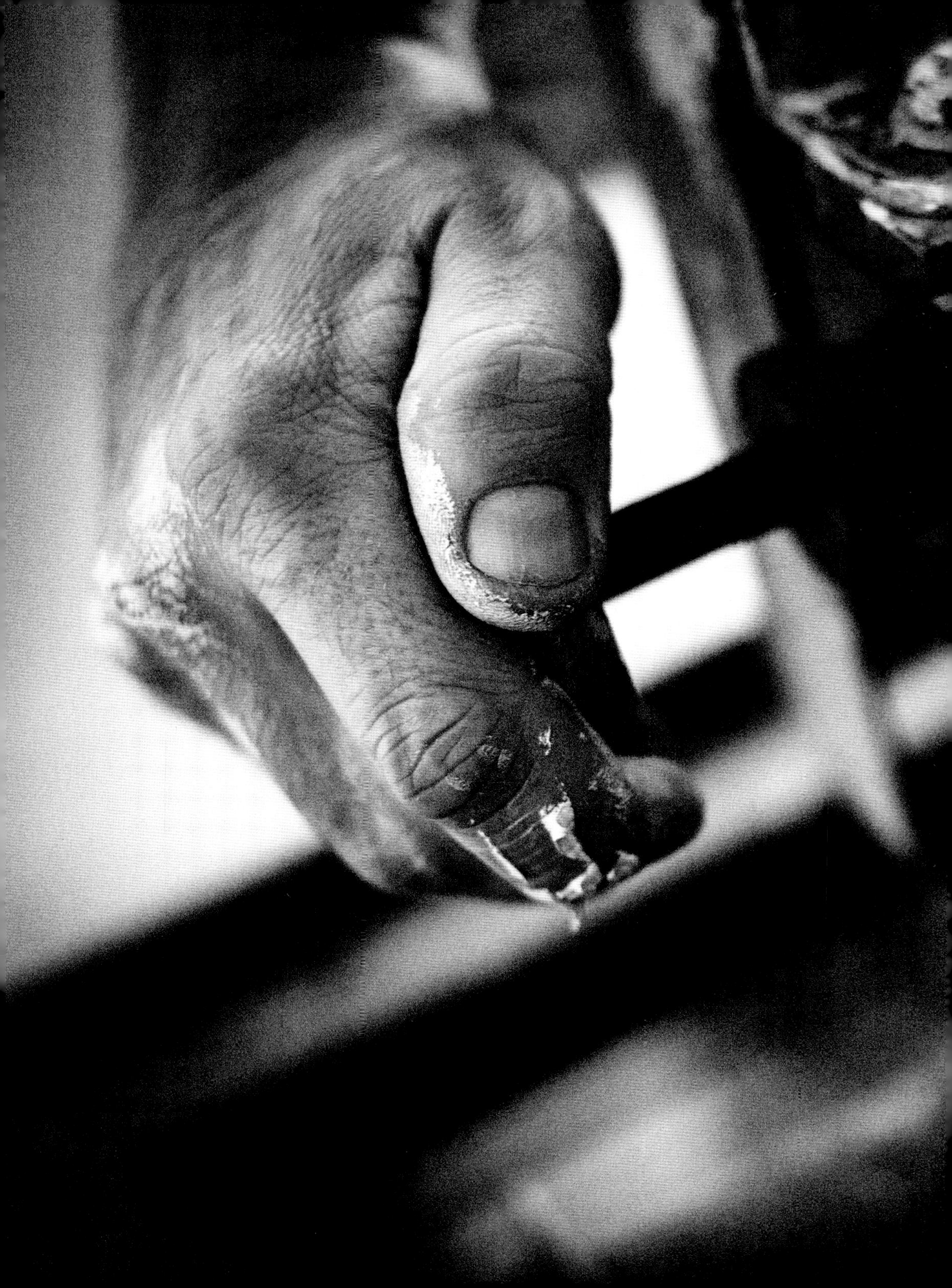

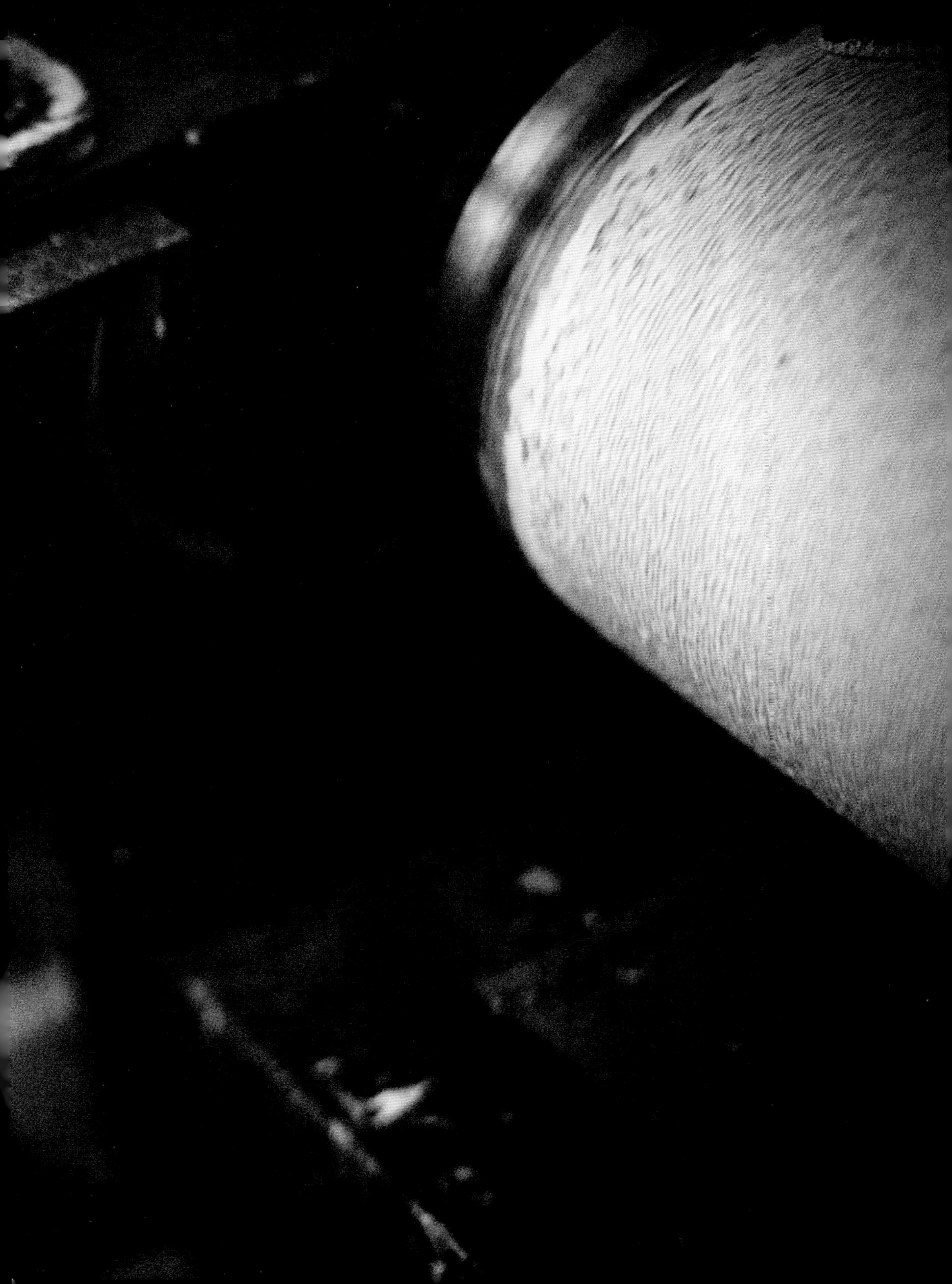

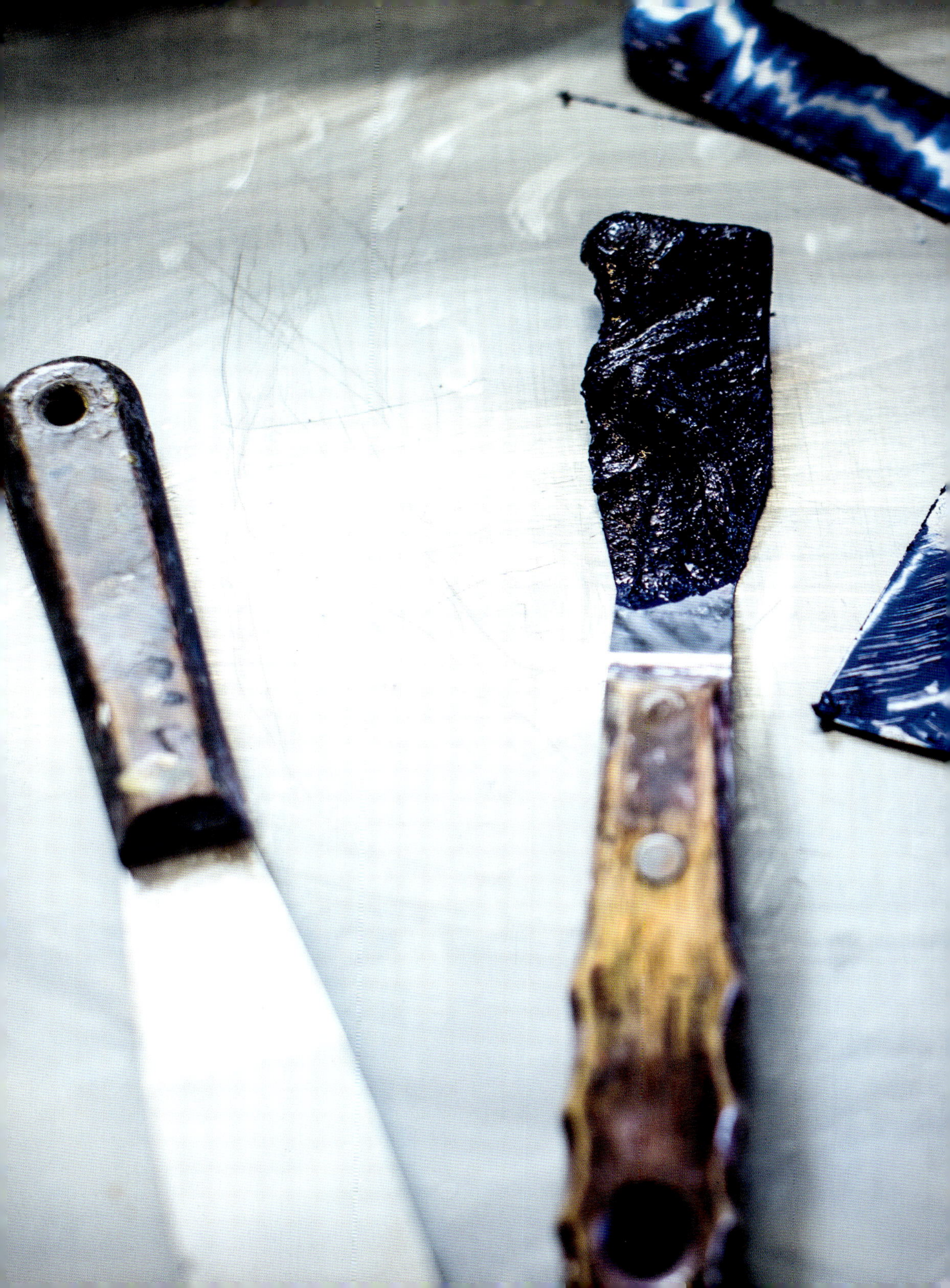

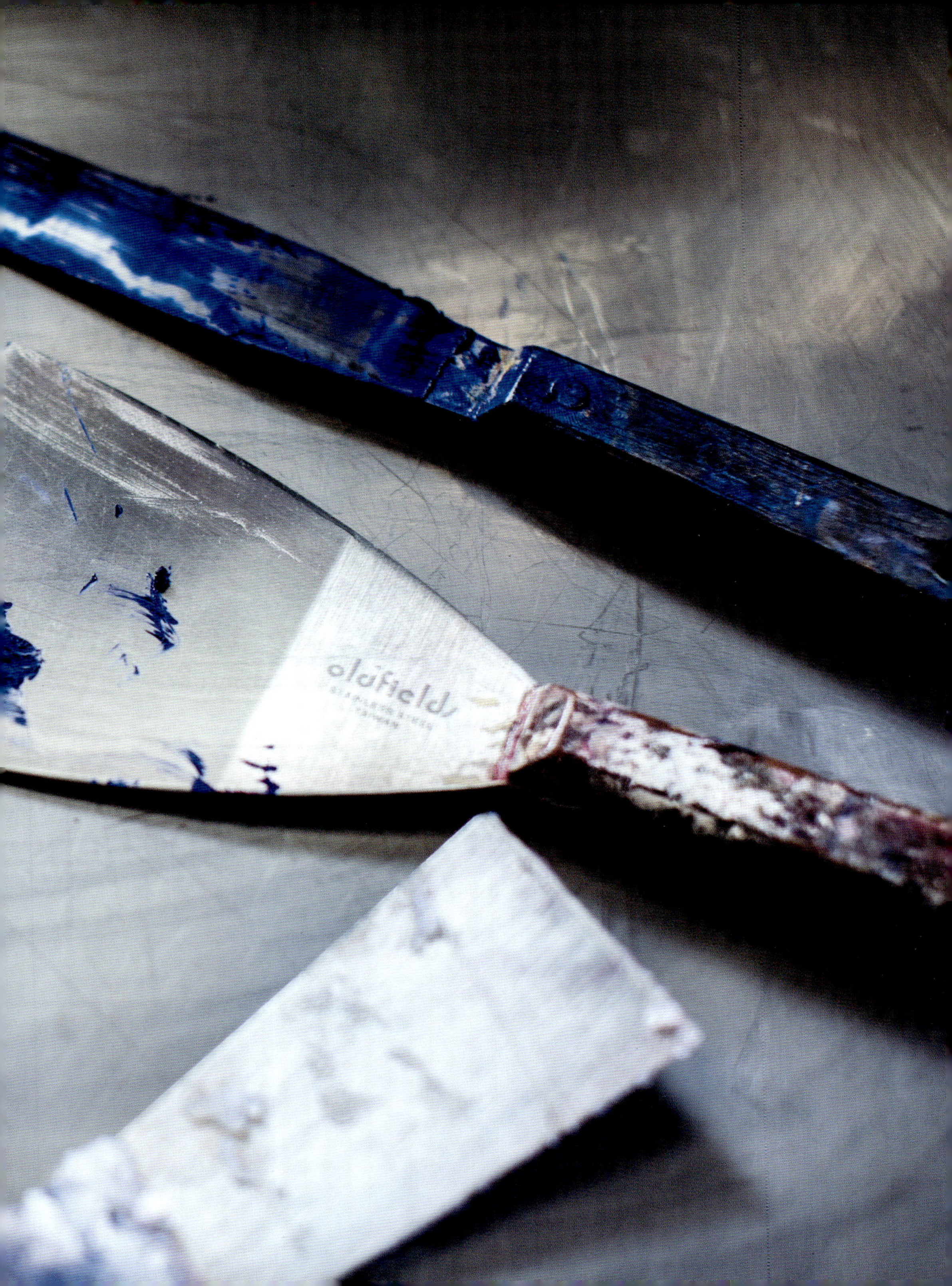
oldfields

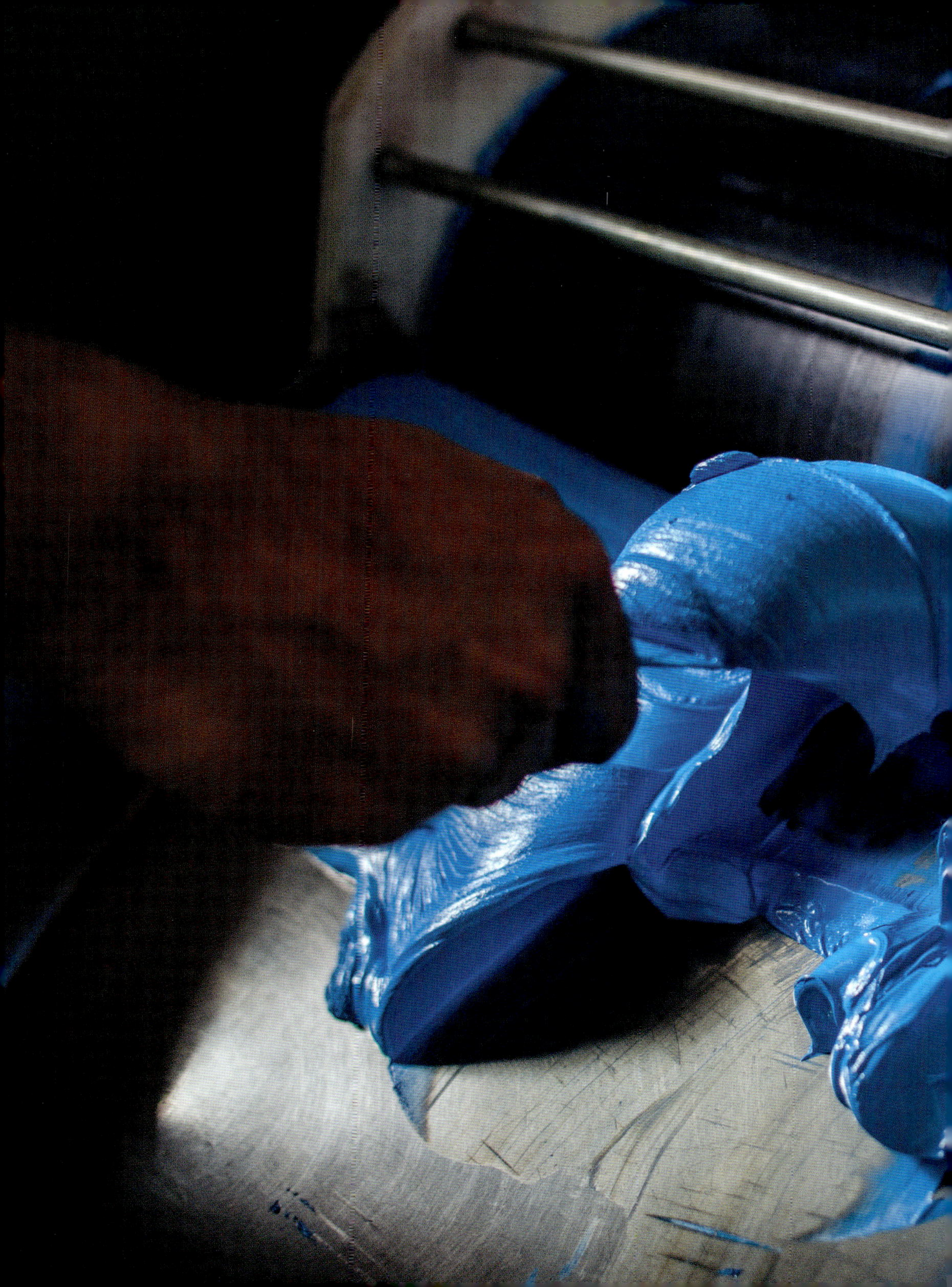

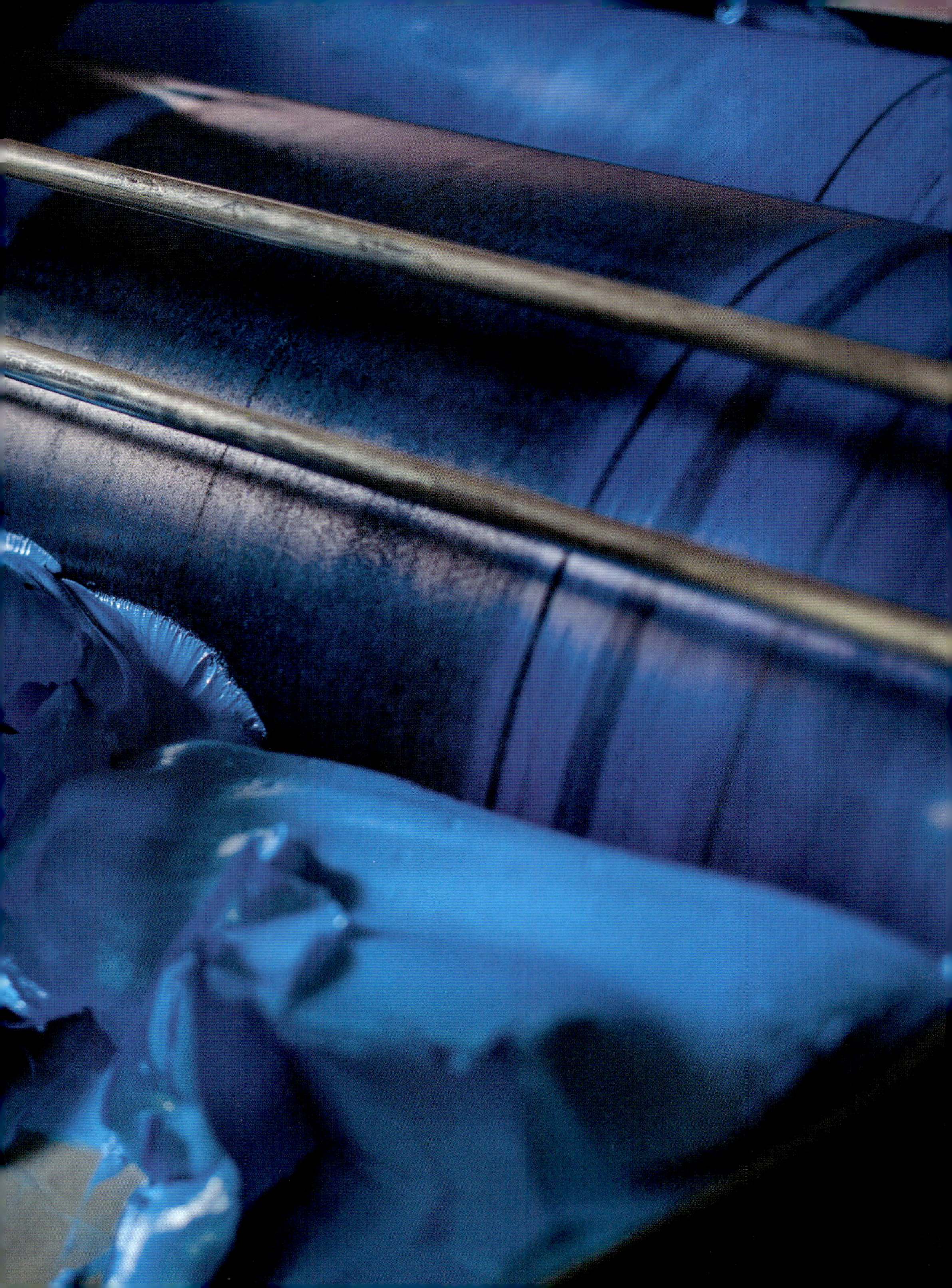

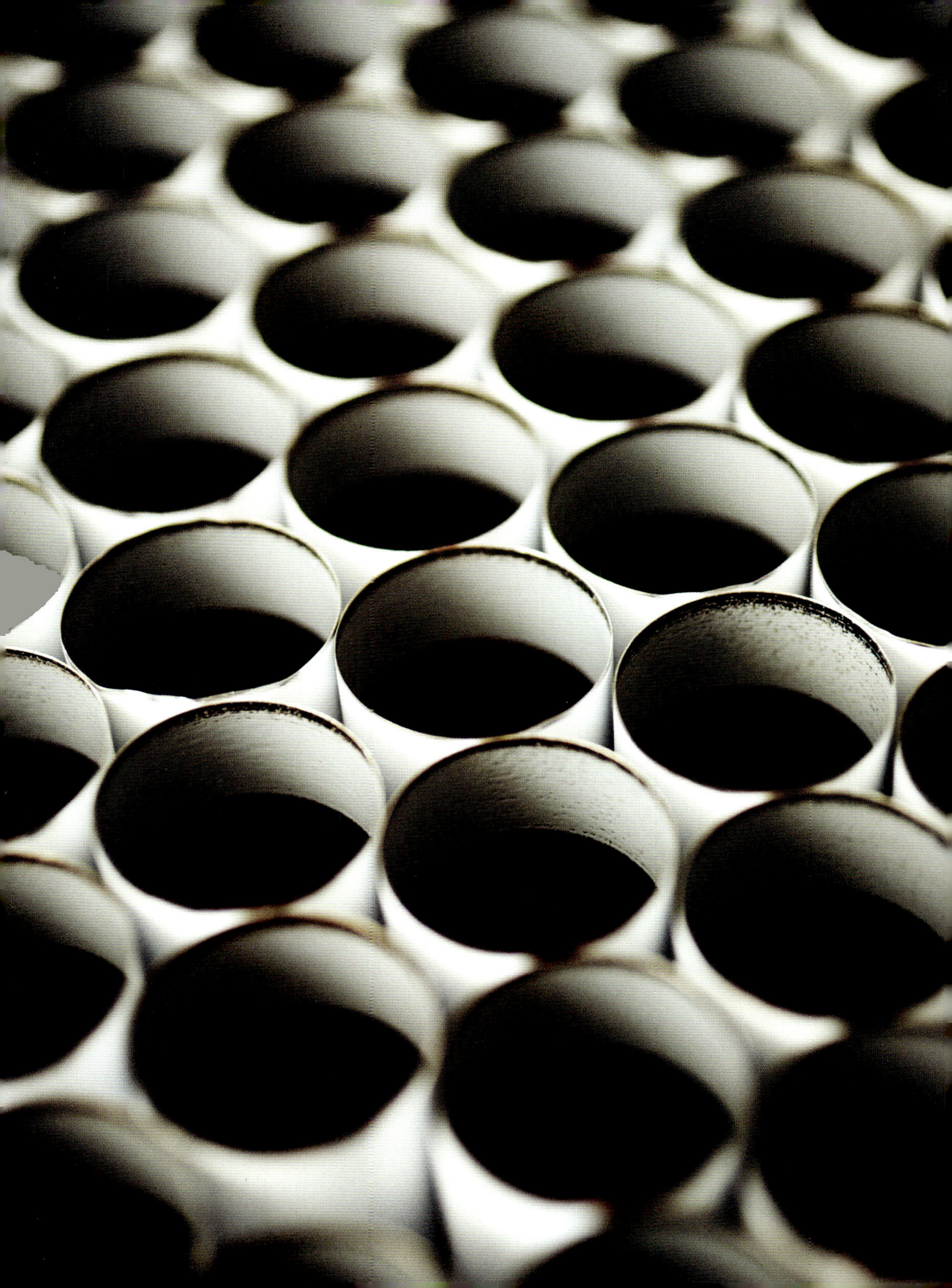

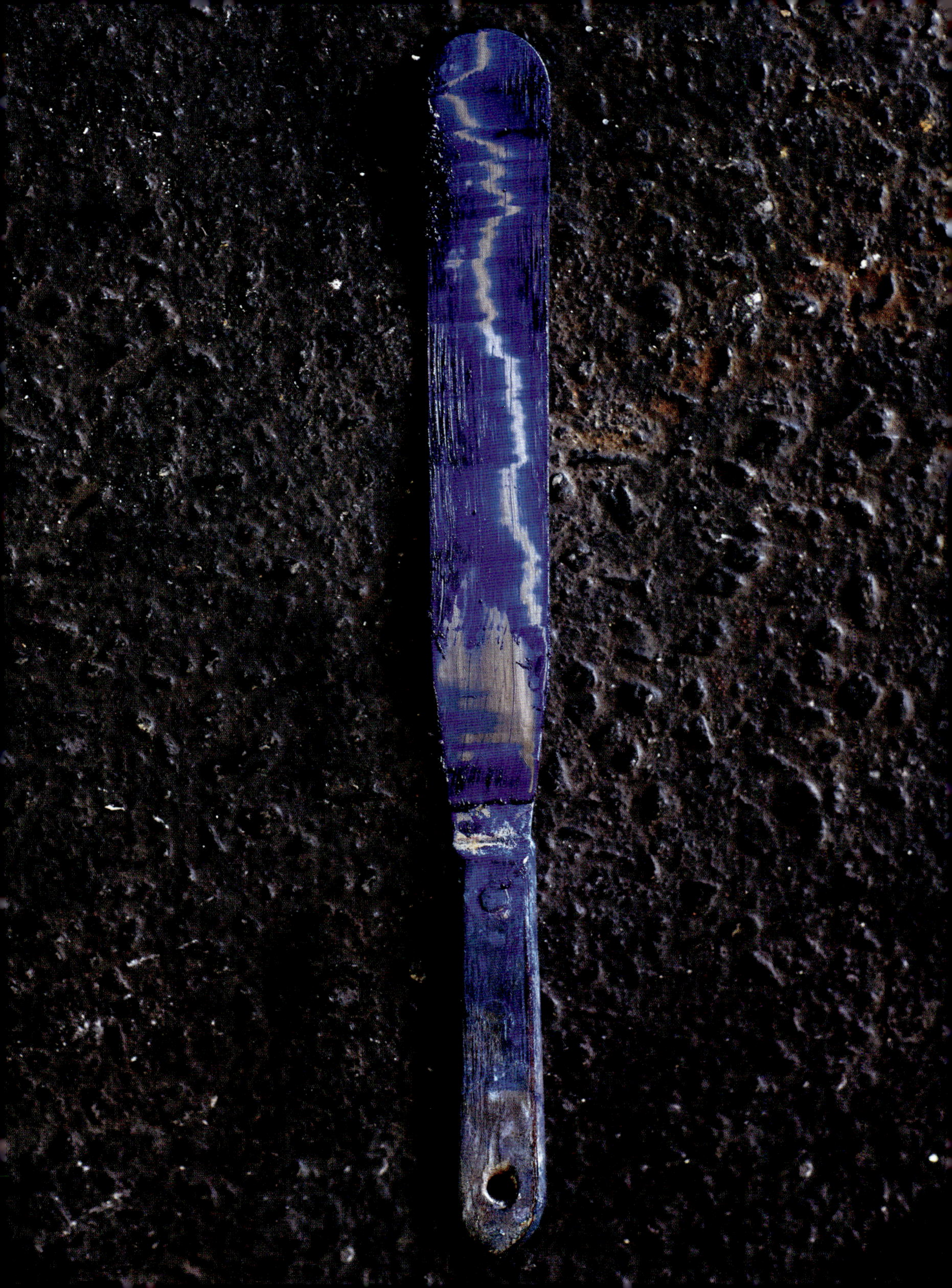

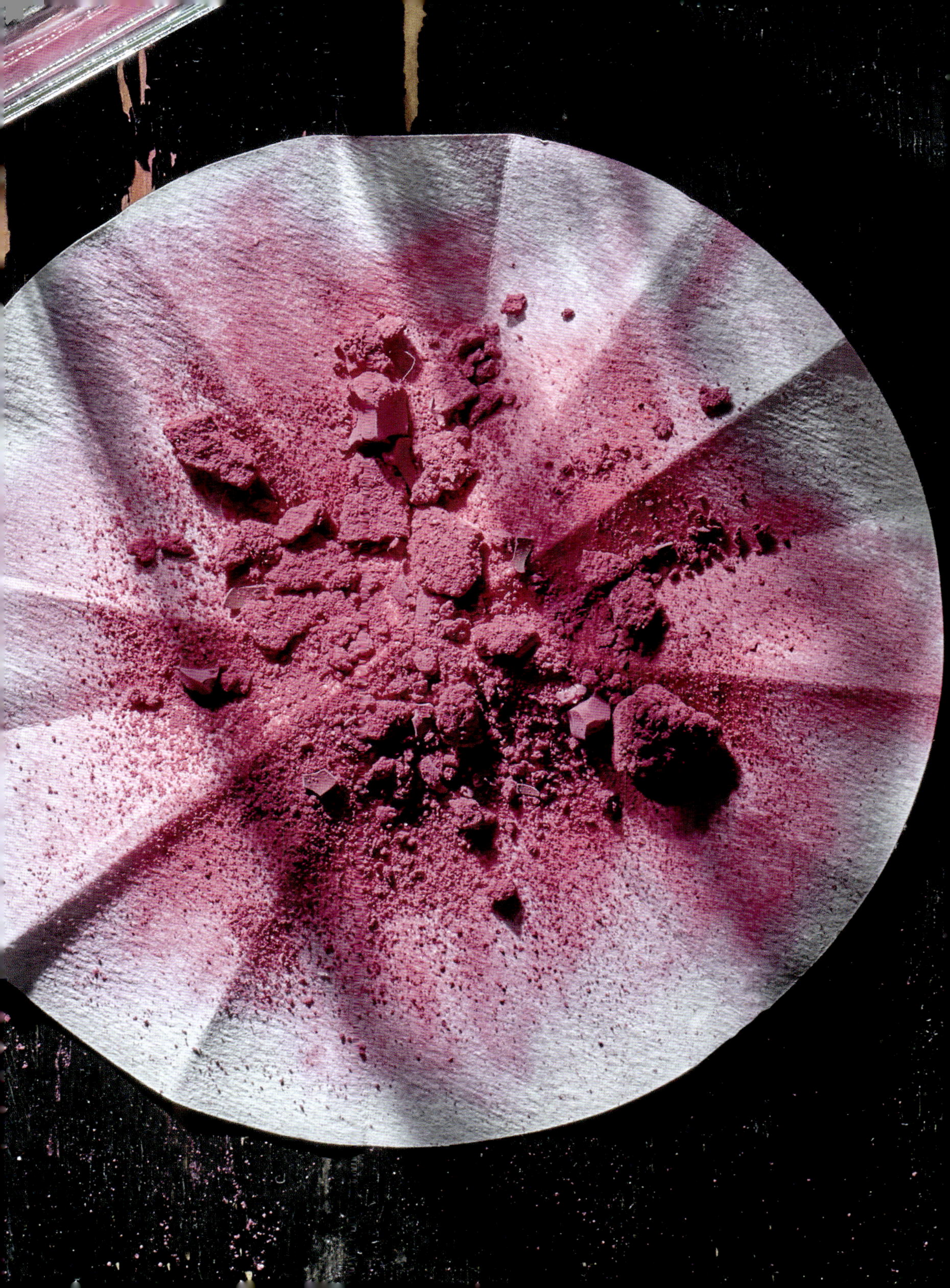

MANUFACTURING PIGMENTS

The following recipes for the manufacture of artists' pigments help explain the processes undertaken. Perfecting the quality of the final pigment involves repeated experimentation. If you wish to make your own pigments, always buy the best grades of raw material possible to ensure a successful result.

LEAD WHITE

Lead sheet	0.8 mm thickness
Vinegar (20% solution)	800 ml
Sugar	500 grams
Dried yeast	15 grams

Divide the vinegar evenly between five bowls and place them carefully inside a large lidded container. Leave enough space for a small pail to be added later.

Cut the lead into five sheets (29 cm x 30 cm). Clean and degrease the sheets on both sides with mineral spirits. Roll the sheets into loose coils lengthways and place each inside a tall vessel.

Sit the vessels in the vinegar-filled bowls. The lead should not be in direct contact with the vinegar, but the fumes must be able to reach the metal coils.

Next, make the solution that will produce carbon dioxide. In a small pail, mix the sugar with 1.5 litres of hot water (approximately 45°C) and stir gently until dissolved. In a small cup, dissolve the yeast in some hot water. Add this to the sugar solution and blend them. Carefully place the pail inside the large container and put the lid on. Keep the container in a warm place to promote the ongoing production of fumes.

In the first hour, check for bubbles in the yeast–sugar solution, as this indicates that carbon dioxide is being produced. The vinegar fumes react with the metal to create lead acetate, and the carbon dioxide converts this to basic lead carbonate. Try not to remove the lid too often, as fumes will escape and this will reduce the chemical reaction.

Replace the yeast–sugar solution every seven days and replenish the vinegar every month. Full conversion into lead white takes three months. Using gloves and mask, because lead white is extremely poisonous, break down the fragile coils and wash them in water to remove any impurities. Filter the pigment and leave it to dry in a warm, draught-free space. Finally, grind it in a mortar and pestle to a fine powder.

Caution! Lead is a poisonous material and extreme care should be followed when handling it. Wear protective gloves and a particulate face mask at all times. Keep all materials in a secure place, out of reach of children and animals.

LEAD SHEET

YEAST

SUGAR

VINEGAR

LEAD CORROSION AFTER 3 WEEKS

LEAD FLAKES

LEAD CORROSION AFTER 6 WEEKS

LEAD CORROSION AFTER 3 MONTHS

WASHED AND POWERED LEAD

CARMINE LAKE

Cochineal	100 grams
Alum	200 grams
Cream of tartar	10 grams
Soda ash	170 grams

To make carmine lake, the soluble dye must be extracted from the cochineal and converted into an insoluble pigment. First, take the cochineal insects and grind them down into a fine powder using a mortar and pestle.

Add the powdered cochineal to 1600 ml of boiling water on a stove and simmer gently for one hour. Strain the liquid through filter paper into a 6-litre heatproof container.

Next, dissolve the alum and cream of tartar in 1600 ml of hot water. Pour this warm solution into the dye container and stir to mix fully.

Now dissolve the soda ash in 1600 ml of hot water. Gently pour this solution into the dye container. It will froth immediately. Stirring to mix the solutions together will generate even more of a reaction.

Allow the solution to sit overnight to let the pigment precipitate settle. Using a siphon, draw off the slightly coloured water above the precipitate until there is only about 5 mm left above the pigment surface. Now add 2 litres of cold water, stir and allow the pigment to settle overnight again. Draw off the liquid as before. Repeat washing until the water is colourless after the pigment has settled.

Filter the pigment and leave it to dry in a warm, draught-free space. Finally, grind it in a mortar and pestle to a fine powder.

COCHINEAL

ALUM

SODA ASH

COCHINEAL EXTRACT

GROUND CARMINE

FILTERED CARMINE

ULTRAMARINE

Lapis lazuli (highest quality)	500 grams
Beeswax	125 grams
Gum mastic	125 grams
Colophony (lump rosin)	250 grams
Soda ash (sodium carbonate)	300 grams

To make the brightest genuine ultramarine, the lapis lazuli must be as pure as possible. Lower-quality ore results in a pale grey-blue pigment.

Break down the ore into gravel with a hammer. Next, grind it to a fine powder using a covered mortar and pestle to prevent stray grains escaping the vessel. Finally, wet the powder and finely mill with a muller and slab. Dry the powder and pass it through a 50-micron sieve.

Melt the beeswax, mastic resin and colophony in a pot until combined. Add the lapis powder and stir until it is fully incorporated. Pour the paste onto a non-stick surface and knead it until the putty is a uniform consistency. Leave it to rest for three days.

Fill three deep-sided bowls each with 4 litres of water heated to 40°C to make your lye baths. Dissolve 100 grams of soda ash in each bowl. Wearing vinyl gloves, or coating your hands with linseed oil, submerge the putty in the first lye bath and massage it. The lazurite contained in the rock is hydrophilic (water-loving) and the other materials are hydrophobic. Massaging the putty underwater causes the lazurite to pass into the water. The impurities remain in the putty.

The best quality blue pigment will leach out of the putty after two hours of massaging. Move the putty into a fresh lye bath and repeat the massage process for one hour, swapping to the final bowl for one final hour of massage. Each bowl will now contain different grades of colour. The first will be deep blue, and the others will decrease in intensity.

Allow the pigment to settle overnight and drain off the clear water. Pour on boiling water to remove any wax-resin residue. Leave to settle again (this can take at least three days, although it can vary), drain and allow to dry.

Using this traditional method will give you approximately 20 grams of the best quality ultramarine pigment.

LAPIS LAZULI ORE

BEESWAX   GUM MASTIC

COLOPHONY

LAPIS EXTRACT IN LYE BATH

LAPIS LAZULI PUTTY

SODA ASH   LINSEED OIL

FIRST GRADE
ULTRAMARINE

SECOND GRADE
ULTRAMARINE

ULTRAMARINE
ASH

MADDER LAKE

Madder roots	300 grams
Potash (potassium carbonate)	100 grams
Alum	100 grams

Dissolve the potash in 4 litres of water and place the loose madder root directly into this solution. Warm the solution to 40–45°C, but do not overheat. Once the solution has reached this temperature, maintain it for the next 6 hours, stirring vigorously from time to time.

Next, remove the madder roots by straining the liquid through a filter bag and into a large plastic bucket. Do not squeeze the bag, as any sediment from the madder root will muddy the final colour.

Dissolve the alum in 2 litres of warm water and slowly add this solution to the madder-dyed liquid, stirring continuously, at which point the mixture should froth. The warmer the solution is (up to 40°C), the more it will froth. Leave this muddy mixture to stand for three days, stirring once or twice a day, just to start it bubbling again.

Add fresh cold water to the solution until the bucket is almost full. Allow to stand undisturbed for 36 hours. This will ensure that most of the coloured pigment sediment will have settled to the bottom.

Siphon off half the liquid, even if this has a red colour, as it contains soluble dyes that are not needed in the pigment. Refill the bucket with fresh cold water and leave it undisturbed for 24 hours. Next, siphon off three-quarters of the water and refill the bucket with fresh water. Continue this washing process until the water is colourless. Allow at least 8 hours between washes, as this is the minimum time for proper settling of the pigment.

After filtering the pigment, allow it to dry in a warm, draught-free space, then grind it to a fine powder with a mortar and pestle.

This recipe extracts the alizarin dyestuff for making madder lake, of which alizarin crimson is the most refined version.

An alternative method of extraction of Pseudopurpurin dyestuff from the Madder root by fermentation, creates Rose Madder.

MADDER ROOTS

ALUM

POTASH

MACERATED MADDER ROOTS

PSEUDOPURPURIN EXTRACT

ALIZARIN CRIMSON PIGMENT

FILTERED ROSE MADDER

ROSE MADDER PIGMENT

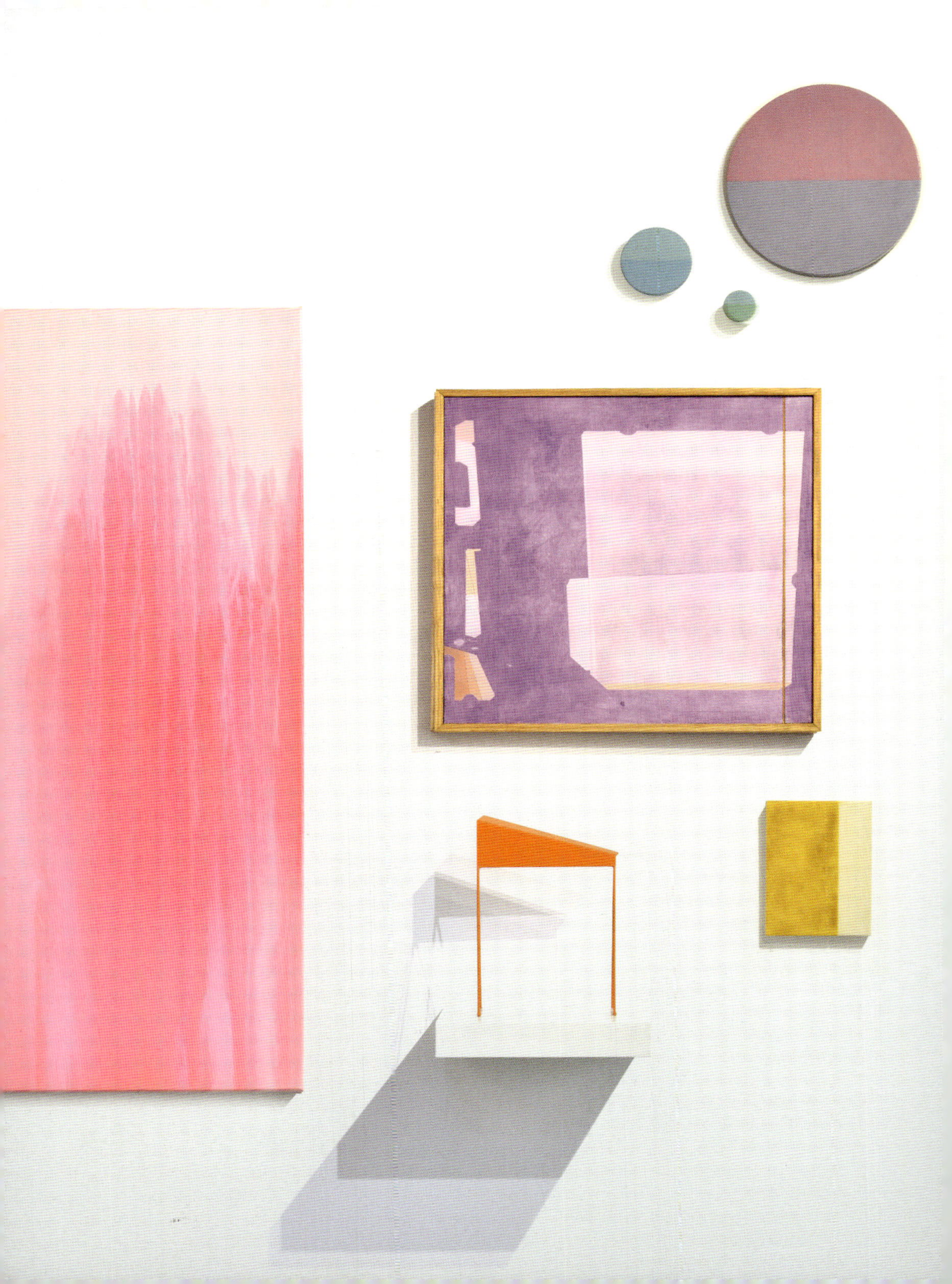

ARTISTS' COLOUR

Pigments have no intrinsic value. It is not until they go through the transformative process of becoming colour and in turn art that their potential becomes realised.

The artists featured in this section celebrate individual colours – and the artworks reproduced here show how the use of paint, pigment, dye or chemical reaction continue the glorious journey of discovering colour.

'Magenta is a rich and beautiful colour I enjoy working with.'

ARTISTS' COLOUR

ASH KEATING
Australia

Gravity System Response #38
Synthetic polymer on linen
2017

188

'Purple is the colour of shadows, that is where it exists.'

IAN WELLS
Australia

Err
Oil and acrylic on traditional gesso
2017

'Blue is not at all mercurial or capricious. It is primary, a wellspring for many colours. It persists and wields influence whether we perceive it or not.'

CONNIE GOLDMAN
United States of America

Untitled
Acrylic on wood
2007

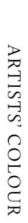

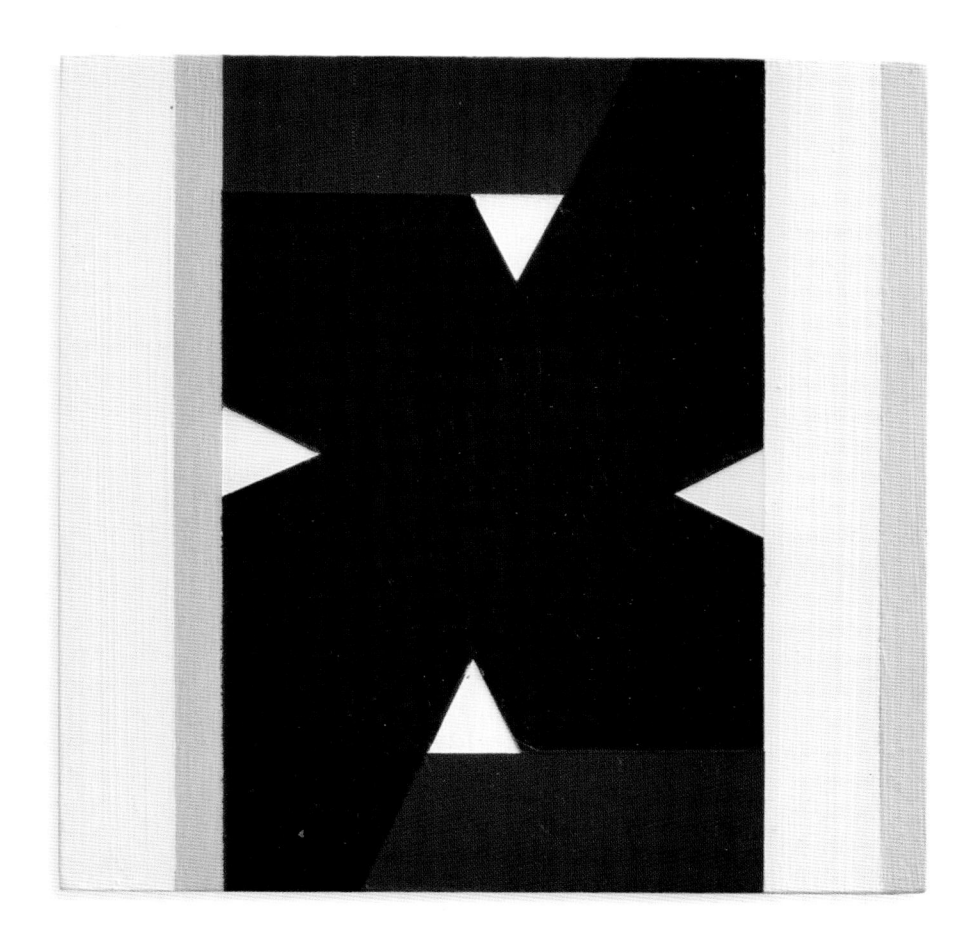

'Black is fantastic; it doesn't say much, hangs in the background. But when it does speak up it does so with a quick wit and is always on point.'

DON VOISINE
United States of America

3-C
Oil on wood
2014

'My work is systems-based and conceptually driven.
I want to portray quality by showing its colour, not its form.'

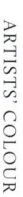

DEBRA RAMSAY
United States of America

Apple in 13 Colors
Acrylic on Dura-Lar
2014

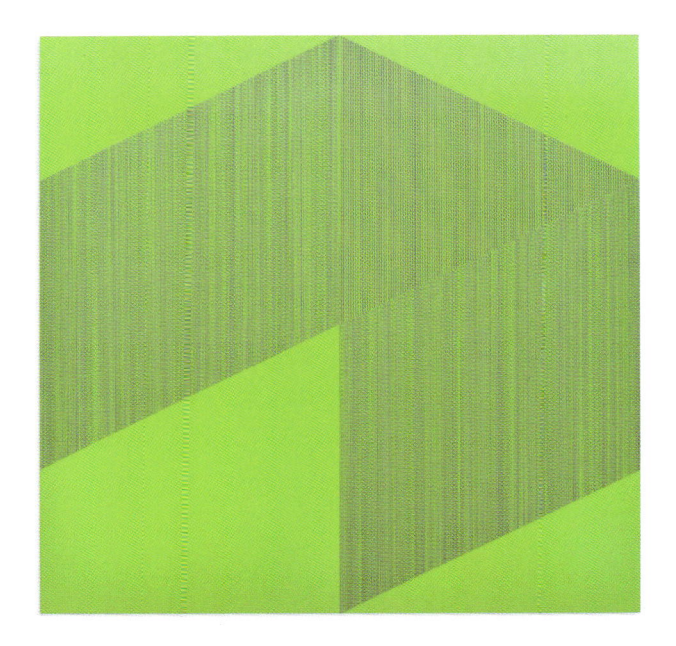

'I tend to select colours spontaneously and this green appealed because – taken by itself and applied over a large area – it feels a little toxic and isn't particularly pretty.'

EMMA LANGRIDGE
Australia

One of two forms
Enamel and acrylic on aluminium
2012

'Working with colour relationships is an aesthetic exercise for me. I'm not thinking about meaning. But since you asked, how about "yellow green is an evocation of the first shoots of spring".'

JOANNE MATTERA
United States of America

Silk Road 207
Encaustic on panel
2014

'I take the position that specific pigments can cause the viewer to reconsider their notion of a given colour. In this case, I had personally prepared a paint with a relatively rare pigment: cobalt blue light. I wanted a pale yet intense blue, referring to the blue veils seen traditionally in paintings of the Virgin Mary.'

KEVIN FINKLEA
United States of America

Parakeet for Palermo
Acrylic on wood
2012

'In choosing to work with white – my intention was to emphasise the form of the support. The range of tonal value, which is afforded by the play of light on the white surface, serves this end. I also wanted to avoid the complication of any emotional response to a particular hue.'

RICHARD VAN DER AA
New Zealand

Mere Formality
Acrylic on wood
2010

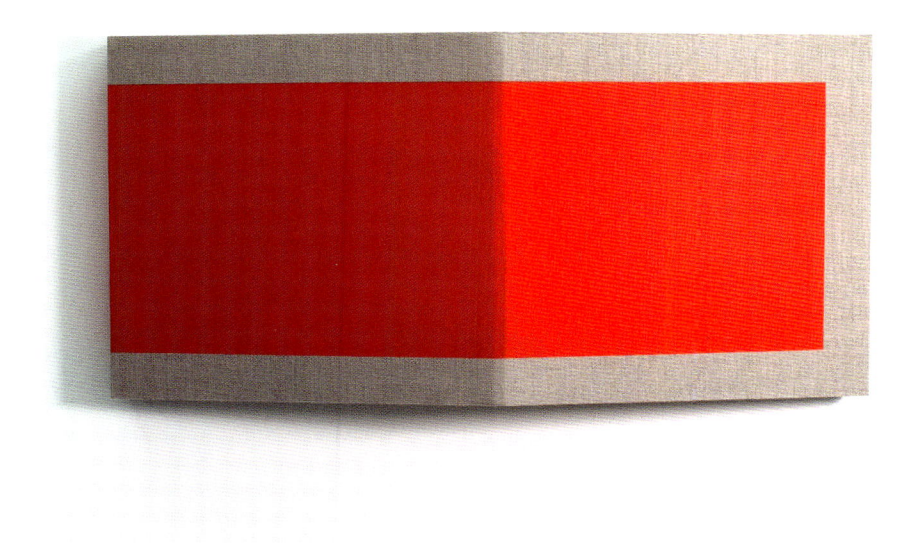

'The composition of [my work] dictates the colour. It will tell me if it needs to be quiet, loud, formal or whimsical. This piece demanded a bold cadmium red.'

LOUISE BLYTON
Australia

When my heart was volcanic 2
Acrylic on linen
2018

'You either love or hate this pink. Pink has to be the right pink, it has to have this capacity.'

IRENE BARBERIS
Australia

Cross Form
Fluorescent PVC acrylic
2016

'Black may be perceived as the absence of colour. Its light-absorbing qualities offer intensity and focus, suggesting depth and timelessness all the while being intangible and elusive.'

SUZIE IDIENS
Australia

Untitled (black pair)
Acrylic on wood
2013

'The essence of graphite for me, in its true representation, evokes mystery. Its dark, elegant, dirty silver is graphite's purity. Resulting in intoxicatingly sexy encaustic paintings.'

MICHELANGELO RUSSO
Australia

Untitled
Encaustic wax on timber
2007

'Yellow triggers a memory of daybreak on a summer's morning a long time ago. It reminds me of the promise of the day. Today, it is the colour of nostalgia and hope.'

MUNIRA NAQUI
United States of America

Harvest Moon
Encaustic on wood
2017

'It's a struggle to give yellow a form as it reads as light, transparent and radiating. It's a squeeze of lemon; a soft, spreadable knob of butter.'

BRENT HALLARD
Australia

Bone Yellow
Acrylic on aluminium
2012

'My studio looked onto a hill with this incline, while the colour was reflecting the burnt Australian landscape during the heat of the summer fire season. I painted the same shape green for winter.'

PETER D. COLE
Australia

Elemental Landscape
Metal
2009

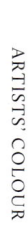

'Iridescent colours are characteristically unstable and interactive. Reflective pigments are highly responsive to physical changes in the surrounding environment, including atmospheric shifts, which can severely adjust the appearance of an artwork in paradoxical ways.'

SAMARA ADAMSON-PINCZWESKI
Australia

Torque 5
Acrylic and iridescent acrylic
on wood panel
2016

'Green is the colour of contemplation. When I find myself alone in the natural environment, free from worldly concerns, it is then that I discover an undeniable sense of tranquillity and peacefulness.'

PETER SUMMERS
Australia

Restrained Directness
Oil on linen
2015

'Piet Mondrian hated green. He called it the colour of decay. So, naturally, I sometimes redress the injustice by insisting on using green in my painting.'

TOM LOVEDAY
Australia

Edge of the World 1
Acrylic on canvas
2012

'The form of the sculpture tells me what colour it will be. Sometimes I ask it: "What colour are you?" In this case, horizontal form and wavy grain in the ash veneer equals oceanic.'

RICHARD BOTTWIN
United States of America

Small Forward
Acrylic on wood
2013

ARTISTS' COLOUR

'I worked with micaceous iron oxide after I heard the [2016 Presidential] election results. It's a dark colour with a lot of depth to express my disdain for the new regime.'

RUTH HILLER
United States of America

Untitled
Acrylic on Dura-Lar
2016

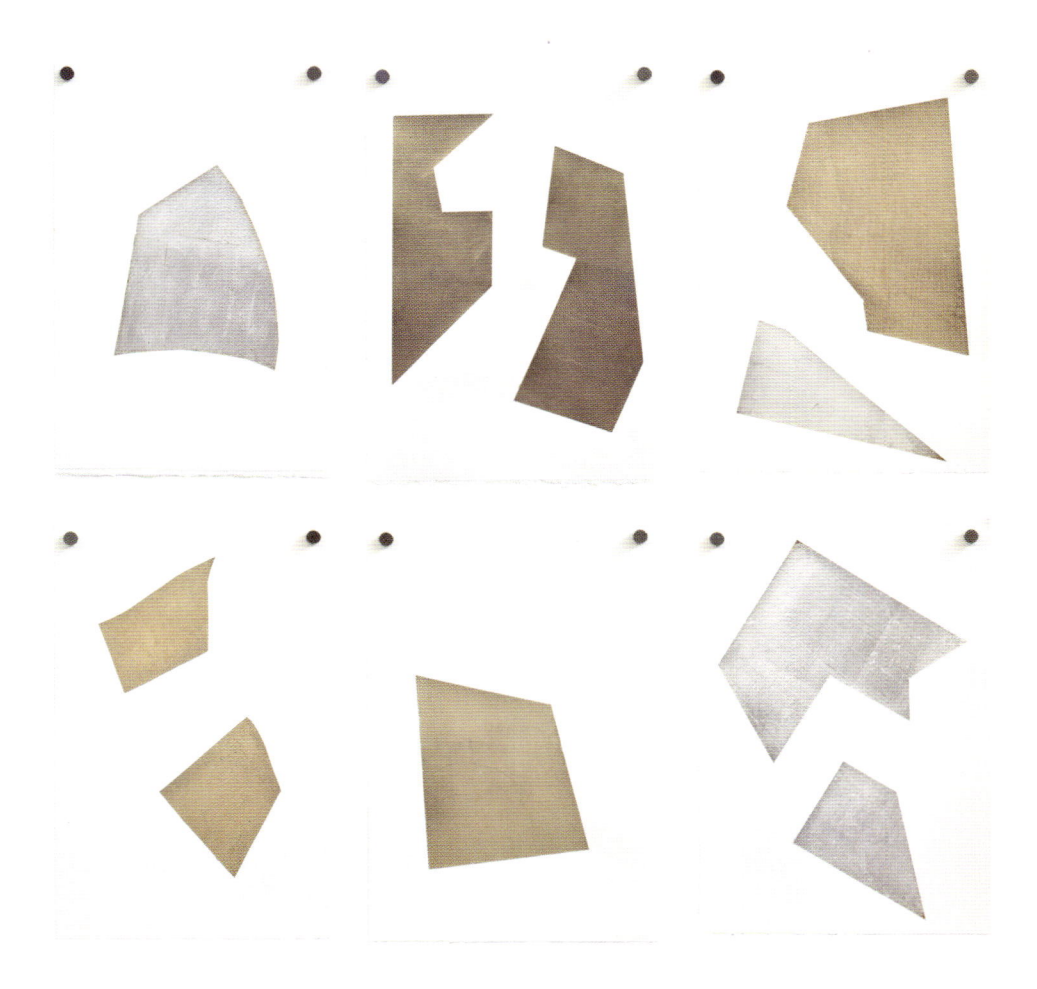

'The reflective lustre of pure gold leaf is captivatingly beautiful. Its soft glow bathes all it surrounds with ethereal luminosity. Medieval painters and illustrators of illuminated manuscripts knew this: I now know it too.'

WILMA TABACCO
Australia

Test Flight No.'s 1-6
Gold Leaf on Paper
2009

'Once rendered in black, the potential for narrative is undermined and not only allows but also invites abstraction into the equation.'

EUAN HENG
Australia

Fall (maquette)
Cardboard and black spray paint
2017

'Black is the colour of night between the stars. It is seductive and sensual and hits hard. It's a familiar stranger, who is always quietly present, even if it's just behind the curtain of light.'

JAMES AUSTIN MURRAY
United States of America

All the Stars in the Sky
Oil on canvas
2018

'If I could only work with one colour for the rest of my life, without question my "desert-island colour" would be indigo. I know it's cheating – like calling the collected works of Shakespeare a single book – because indigo contains such a range of shades and subtleties depending on the number of dips and the formulation of the dye bath.'

JEANNE HEIFETZ
United States of America

Mottainai 11
Ink on Gampi Torinoko paper,
hand-dyed with indigo
2016

'I was once advised by an art dealer to never go too large or paint nudes and to never, under any circumstances use more than 20% green in any composition ... all very hard to sell he complained. I've been applying a verdigris finish to everything ever since!'

SIMON LEAH
Australia

Untitled
Copper, silicone,
Victorian oak and pine
2017

'Red is a strong colour that can evoke a certain mood, create a message or a sharp response. It is the colour of assertion, strength and vitality. It often signifies a warning. This industrial colour creates a dichotomy between attraction and caution.'

MARLENE SARROFF
Australia

Illuminous Red
PVA strips
2014

'If all my colours were taken away from me leaving me with only Cremnitz White No. 2 (ground in walnut oil) I'd still be able to get up to no good.'

JIM THALASSOUDIS
Australia

The KKK Took my Colours Away
Lead white oil paint, wood and glass
2017

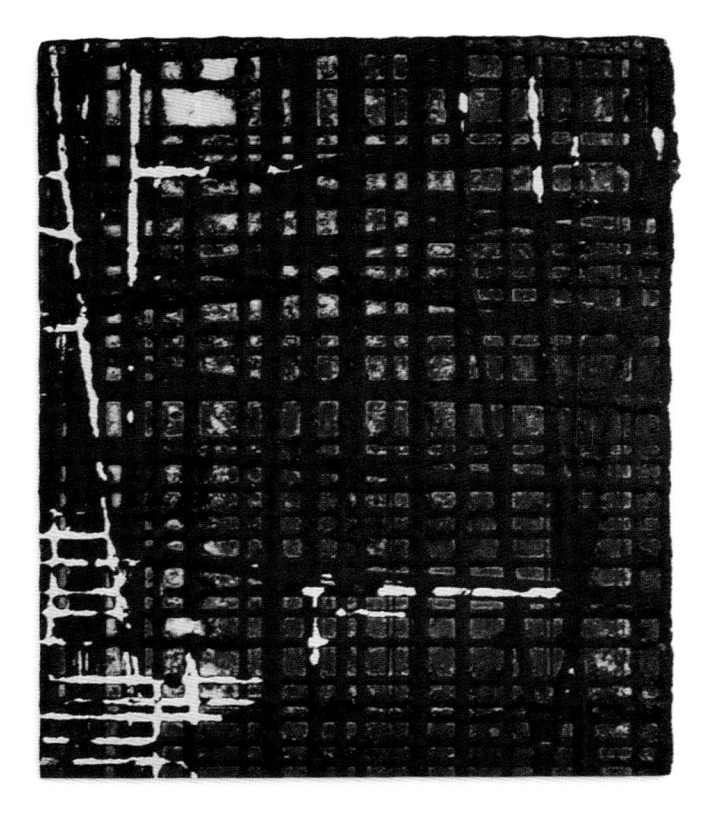

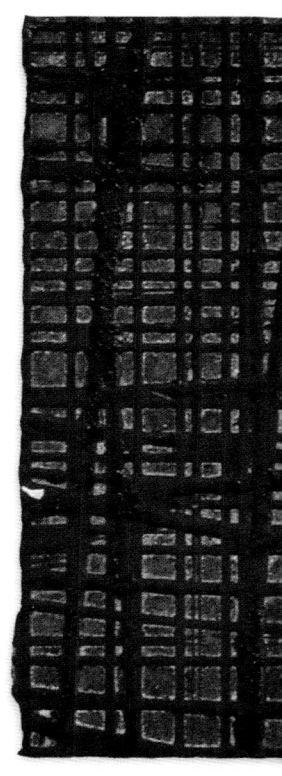

'Whilst devoid of hue, colour references can be drawn from within the deep blackness and patina present upon the surface of the ink.'

TJ BATESON
Australia

Iteration Black Linear Structure I, II, III
Woodblock engraving
2017

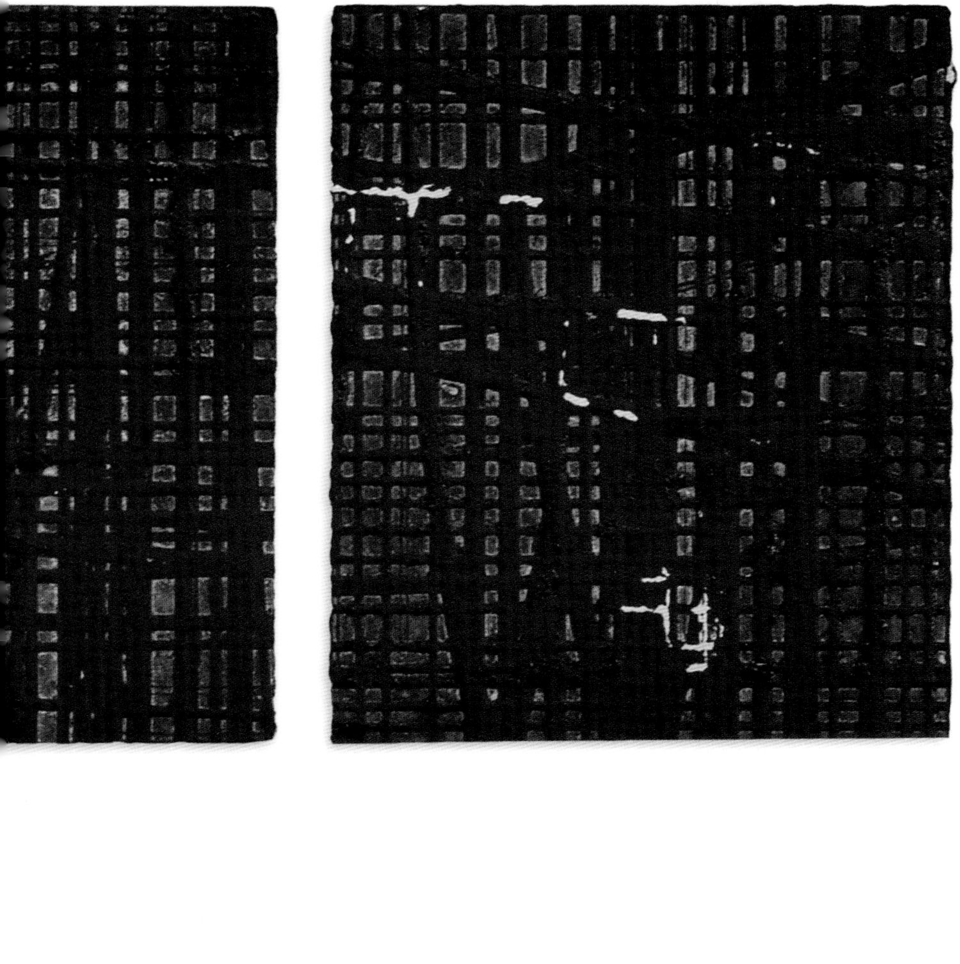

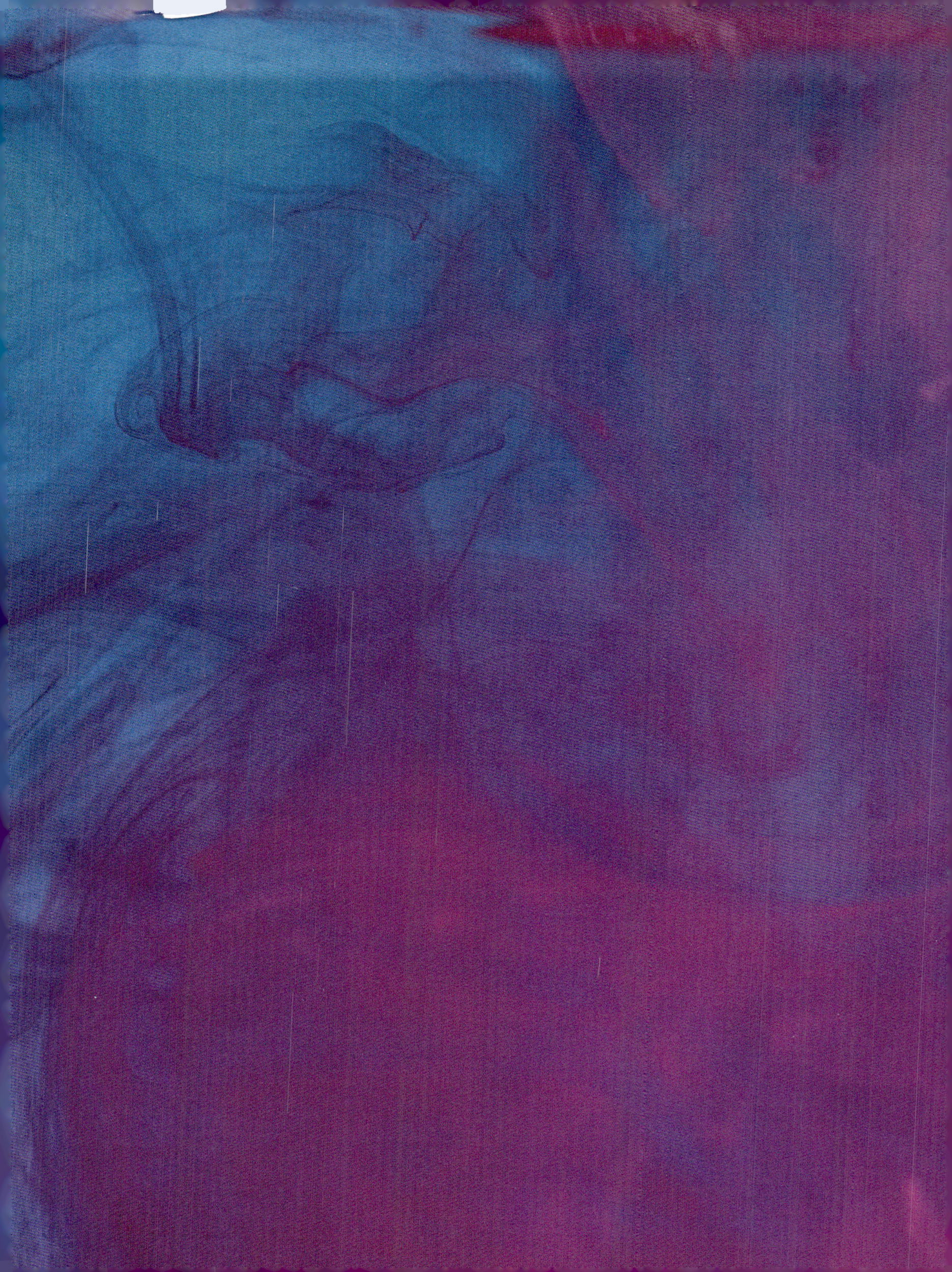

ENDNOTES

1 Cennino Cennini, *The Craftsman's Handbook*, translated by DV Thompson. Dover Books, 1932.
2 Cited in Andrew Dalby, *Dangerous Tastes: The Story of Spices*. University of California Press, 2002.
3 Cennino Cennini, op. cit.
4 Aristotle, *Parts of Animals*, translated by Arthur Leslie Peck and Edward Seymour Forster. Harvard University Press, 1937.
5 William Dampier, *A New Voyage Round the World*. Dover Books, 1969.
6 *The Travels of John Sanderson in the Levant (1584–1602)*, edited by William Foster. The Hakluyt Society, 1930.

ACKNOWLEDGEMENTS

I would like to thank the many people who have made this book possible.

To my beautiful wife, Louise, for her unwavering support of my wanderlust over the years. You are my true co-conspirator without whom nothing in the last twenty years could have been achieved; *Chromatopia* is truly a joint effort with you.

To Adrian Lander, photographer extraordinaire, for turning this book into something way beyond my original ambitions and expectations. You have brought an enthusiasm and dedication to this project that turned it into a *tour de force* of visual decadence.

To Tahlia Anderson at Thames & Hudson for your curiosity in the original idea, constant belief in this project and steady encouragement and professional astuteness in wrangling this book into reality. Also to the book's unsung heroes: Lorna Hendry for asking the hard questions in manuscript editing and Evi O for the fabulously original design aesthetic.

On the Langridge home front: to Brendan Byatt, whose unflappable professionalism has allowed me so much time away from the business to write this book.

Honourable mentions go to all those who were involved with the original exhibition that spawned this book: Keith Lawrence and Tim Bateson, directors of Tacit Gallery, for having the vision; Ian Wells; Luke Pither; Julie Keating; and Rosalind Atkins whose help, beyond the call of duty, must be recognised.

On the technical side: to Narayan Kandakhar of the Forbes Collection for his generosity of information towards this project; to Keith Edwards for sharing his knowledge in the craft of pigment manufacture; as well as John Hewiston, Nicholas Walt, Andrea Dolci and John Downie.

Finally, to my father, Tony Coles, without whose encouragement to be curious and willing to follow my heart I would never have had such a fulfilling career.

First published in the United Kingdom in 2018
by Thames & Hudson Ltd
181A High Holborn, London WC1V 7QX

This paperback edition 2020

Chromatopia: An Illustrated History of Colour © 2018
Thames & Hudson Australia Pty Ltd
Text © 2018 David Coles
Photographs © 2018 Adrian Lander

p 106: José Antonio de Alzate y Ramirez, *Indio que recoge la Cochinilla con una colita de Venado [Indian collecting cochineal with a deer tail]* (1777). Coloured pigment on vellum. Collection: The Newberry, Chicago. Courtesy of Wikimedia Commons.

p 107: Illustration from 'An account of the Polish cochineal: In a letter to Mr. Henry Baker, F. R. S. from Dr. Wolfe, of Warsaw'. *Philosophical Transactions*, Volume 54, 91–98, published 1 January 1764. Engraving by J. Mynde. Courtesy of Wikimedia Commons.

p 109: J. Pass, *Cochineal cactus (Nopalea cochenillifera) with insects that feed on it, including the cochineal insect (Dactylopius coccus)* (c. 1801). Etching after J. Ihle. Platemark 24.1 x 19 cm. Collection: Wellcome Library no. 25390i. Courtesy of Wikimedia Commons.

Page 186–206, 208–212, 214, 217–219: Courtesy of Langridge Artist Colours

Page 207: Courtesy of Tom Loveday

Page 213: Courtesy of James Austin Murray

Page 215: Courtesy of Simon Leah

Page 216: Courtesy of Marlene Sarroff

The moral right of the author has been asserted.

All rights reserved. No part of this publication may be reproduced or transmitted in any form or by any means, electronic or mechanical, including photocopy, recording or any other information storage or retrieval system, without prior permission in writing from the publisher.

British Library Cataloguing-in-Publication Data
A catalogue record for this book is available from the British Library

ISBN 978-1-760-76121-9

Design: Evi O. Studio / Evi O.
Editing: Lorna Hendry
Cover: Adrian Lander
Printed and bound in China by RR Donnelley

FSC® is dedicated to the promotion of responsible forest management worldwide. This book is made of material from FSC®-certified forests and other controlled sources.

MIX
Paper from responsible sources
FSC® C144853

To find out about our publications,
please visit **www.thamesandhudson.com**.
There you can subscribe to our e-newsletter, browse or download our current catalogue, and buy any titles that are in print.